## World University Library

The World University Library is an international series
of books, each of which has been specially commissioned.
The authors are leading scientists and scholars from all over
the world who, in an age of increasing specialisation, see the
need for a broad, up-to-date presentation of their subject.
The aim is to provide authoritative introductory books for
university students which will be of interest also to the general
reader. The series is published in Britain, France, Germany,
Holland, Italy, Spain, Sweden and the United States.

*Frontispiece* A herd of adult iguanodons, displayed in the Brussels
Royal Museum of Natural History.

Björn Kurtén

# The Age of the Dinosaurs

World University Library

Weidenfeld and Nicolson
5 Winsley Street London W1

For Joachim

© Björn Kurtén 1968
Phototypeset by BAS Printers Limited, Wallop, Hampshire
Printed by Officine Grafiche Arnoldo Mondadori, Verona

# Contents

| | | |
|---|---|---|
| 1 | Theories and discoveries | 7 |
| 2 | Earth history before the dinosaurs | 50 |
| 3 | The dinosaurs | 82 |
| 4 | Other terrestrial vertebrates | 128 |
| 5 | Vertebrates of the sea | 149 |
| 6 | Flora and invertebrates | 169 |
| 7 | Geography of the Jurassic and Cretaceous | 191 |
| 8 | After the dinosaurs | 211 |
| | Appendix: A classification of the Vertebrata of the Jurassic and Cretaceous periods | 241 |
| | Glossary | 246 |
| | Bibliography | 250 |
| | Acknowledgments | 252 |
| | Index | 253 |

# 1 Theories and discoveries

## What is a dinosaur?

Dinosaurs, more than any other creatures of the past, have a popular image. Museum visitors flock to the dinosaur exhibits, whether to get factual information about them, to try to visualise what life was like when these monsters existed in the flesh, or just to wonder at their odd and outlandish shapes. All of these motives are certainly justified: a dinosaur skeleton is a challenge both to the imagination and to the intellect. This may be especially true where the skeletons are exhibited still in the living rock where they were found, as they are at Dinosaur National Monument in Utah and Dinosaur Park in Alberta.

But dinosaurs may also be encountered in cartoons, comics, and light fiction, usually in places where they have no business to be; for instance in the company of cave men, or as surviving monsters on isolated high plateaux in the distant jungle. Dinosaurs may also appear in speculations about the great sea serpent, the Loch Ness monster, and other local monsters.

However, so far as we know, dinosaurs became extinct many million years before the first men appeared. It is true that even now there are scientists ready to contest this statement, as most recently evinced in an interesting if somewhat uneven book, *On the track of unknown animals*, by B. Heuvelmans. But the evidence of surviving dinosaurs is so specious (old reliefs showing dragon-like beasts; sensational tales by travellers unable to document their stories) that it has to be viewed with profound scepticism – at least until somebody, like Professor Challenger in Conan Doyle's story, brings back a living specimen.

It is, of course, sad to think that no human being will ever see a dinosaur in the flesh. It is amusing to picture a dinosaur in a modern landscape. But they do not belong in our world, and would not be able to survive for long; they belong in a landscape that passed away aeons ago, a landscape buried under the surface of the earth, and only resurrected to a kind of part-life in scientific restorations.

Are the dinosaurs, then, merely to be viewed with a mixture of awe and pity, as a form of life that was too old-fashioned and inefficient to survive; as monstrous failures? This may be answered with the simple fact that dinosaurs were the dominant forms of life on land for more than one hundred million years. That is hardly a tale of failure, and it may perhaps even have a lesson for modern dreamers about the Millennium – an ambition that appears very modest in this connexion. Yet we have a potential future, perhaps fully as long as that of the dinosaurs in the beginning of their history.

Though dinosaurs have now been known for more than a century – it was in 1842 that Richard Owen coined the term dinosaurs – research in recent years has continued to add discoveries about them and their world. Most people, influenced by the herculean size of some exhibited skeletons, think mainly of dinosaurs as enormously big and powerful, in fact there were also dinosaurs no larger than a dog. Dinosaur brains are small, and this has caused many students to think that they must have been very dull, sluggish automata. However, as G.L.Jepsen says, brains are like wallets: contents are more important than size. In fact there are reasons to think that the dinosaurs were highly active, motorically efficient animals.

To go into details, the duck-billed dinosaurs have generally been thought to resemble ducks in their mode of life, seeking their food by rooting about on the bottom in shallow water. Yet we now have direct proof that this is wrong, in the shape of fossilised remains of foodstuffs. Many other time-honoured interpretations have also had to be revised.

The word *Dinosauria* has been formed from the Greek words *deinos*, 'terrible', and *sauros*, 'reptile'. The term 'terror reptiles' is indeed very apt for some of these animals. However, Dinosauria is not now used as a classificatory term, for we now know that the dinosaurs belonged to two separate groups of reptiles.

In order more fully to understand the dinosaurs and their history, they should however be viewed in their environment and against the background of the long history of the earth, in which the hundred million years of their reign is but an episode.

## Dinosaurs and systematics

The dinosaurs, then, were reptiles, like the snakes, lizards, crocodiles, and turtles of the present day. A systematist groups all these and their extinct allies into the class Reptilia, belonging to the Vertebrata. Other vertebrate classes are, for instance, Mammalia or the mammals, to which we ourselves belong, and Aves, or the birds.

Zoological systematics is sometimes regarded as a dry-as-dust occupation, devoted to the pigeonholing and labelling of moth-eaten museum specimens. Actually it could be maintained that systematics is more of an art than is any other branch of biological science.

Of course, even to talk about organisms, we need names for them, so that we have to start by making some kind of classification. A more ambitious classifier, however, works on another plane altogether. He attempts to integrate and utilise the total body of information about the organisms studied – morphology, genetics, behaviour, fossil history, etc. – and to weigh all the items in order of significance. His task is then to present a hierarchic scheme giving the degrees of relationship and diversity, and the relative proximity of common descent, within his group. The degree to which individual personality enters in this work is well illustrated by Tate Regan's well-known definition of the zoological species as

> a community, or a number of related communities, whose distinctive morphological characters are, in the opinion of a competent systematist, sufficiently definite to entitle it, or them, to a specific name.

In fact, more objective criteria are available for the species (this will be discussed below, in connexion with the theory of evolution), but Regan's definition retains its validity for the higher systematic categories: the genus, the family, the order, the class, the phylum, and the kingdom. This is so despite interesting recent attempts to quantify morphological and other differences and leave it to the computers to construct a classification. At present, a successful classification still merits description as *un coin de la Création vu à travers un temperament*.

To exemplify some aspects of classification, let us return to the dinosaurs, members of the class Reptilia. Now there are many kinds of reptiles in existence in our time, and if we include the extinct reptiles too, the total variety becomes quite bewildering. Hence we divide the class into a number of subclasses, each of which contains a number of more closely related reptile forms. The snakes and lizards, for instance, are so closely related to each other that they are brought together in a single subclass, the Lepidosauria, while the turtles are more distantly related and are placed in another subclass, the Anapsida, together with some extinct reptiles. The dinosaurs belong to a third subclass with the appropriate name Archosauria, or 'ruling reptiles'.

They are not alone in the Archosauria, however. The ruling reptiles are divided into five different orders, among which may be noted the two orders of dinosaurs, the Saurischia and Ornithischia. The subclass also contains the order Crocodilia with the crocodiles, alligators, caimans, and gavials; the order Pterosauria or 'flying lizards'; and the order Thecodontia, which is a sort of early, basal group in the subclass, out of which the four others have evolved. The order Crocodilia is the only one that now survives.

Now we can see why the term Dinosauria is not used any more in its original sense: this is because the Saurischia and Ornithischia are no more closely related to each other than, for instance, to the Crocodilia.

Orders are further subdivided into families, genera, and species. To permit greater flexibility, intermediate categories may be introduced by the use of prefixes. For instance, the great carnivorous dinosaur *Tyrannosaurus rex* is a species of the genus *Tyrannosaurus*, which belongs to the family Deinodontidae, of the infraorder Carnosauria, suborder Theropoda, order Saurischia, subclass Archosauria, class Reptilia, super-class Tetrapoda, subphylum Vertebrata, phylum Chordata, kingdom Animalia.

Classification, however, does not only serve the purpose of expressing relationships for their own sake. The affinity between dinosaurs and crocodiles has a special significance to us. Of the former there is little more left than the fossil bones, and many of their characters – such as the nervous system, the heart and blood circulation, etc. – cannot be studied directly. In these respects, however, we may assume that the dinosaurs resembled crocodiles more than any other living reptiles, simply because the crocodiles are the nearest living allies of the dinosaurs. Thus a study of crocodiles may help us to understand the dinosaurs better.

The birds are also quite closely related to the dinosaurs. This may sound surprising, but in fact birds have evolved from the same ancestors as dinosaurs. We might indeed play with the notion of living birds as kinds of feathered and winged dinosaurs. Though this might seem fanciful as regards the song birds, anyone who spends some time observing an ostrich, or even a swan, will I think agree. So we cannot exclude the possibility that the characters and behaviour of the birds may also tell us something about the extinct dinosaurs.

## What is a fossil?

How can we know anything at all about dinosaurs and other animals which became extinct long ago? Our sole source of direct knowledge about them are the remains called fossils. The science dedicated to their study is called palaeontology.

The word fossil originally signified something dug out of the earth, and was applied to any object of this kind. It is now restricted to mean traces of organic life only. Nevertheless, there are many kinds of fossils.

Remains of dinosaurs, for instance, are mostly preserved in the form of more or less petrified bone. When a dinosaur died, its carcass might under certain circumstances become covered by

rapidly accumulating sediments. The soft parts would rot away, but bone and teeth would remain longer. Ultimately these, too, may be destroyed, but it may also happen that the bony tissues become impregnated by minerals brought in solution by percolating water. In this way the bones may gradually become petrified, so that their chemical composition is completely altered, while the shape remains unchanged. However, if the sediment is compacted by the weight of other strata accumulating on top of it, the fossil will be flattened too. Other deformations occur as a result of shearing stresses introduced by movements of the earth's crust, for instance in connexion with mountain-building (orogeny).

This would be a fairly typical case history for one of the dinosaur skeletons exhibited in our museums. But petrifaction does not necessarily occur. In some fossils, especially those of more recent age, the original chemical composition of bone or shell may be retained with little change.

Again, the entire skeleton will be preserved only when the whole animal has been imbedded. But more often the dead bodies will be torn apart by predators, carrion-eaters, or simply by the action of running water as the tendons and muscles rot away. A floating,

1 The excavation of a large dinosaur skeleton is
a major operation, requiring technical skill,
time and patience; the bones are often brittle
and have to be painstakingly hardened. Skeleton of a
sauropod dinosaur being excavated in Mongolia,
showing backbone, ribs and (left) limb bone.

putrefying carcass will lose its limbs, its lower jaw, and its skull at different points; and so it may happen that the palaeontologist finds nothing but loose bones and teeth to work on.

In quite exceptional circumstances, soft parts of organisms may become preserved. In a desert, a carcass may dry out and become mummified, perhaps later to be imbedded in silts brought by a seasonal flood; then the remains of the dried hide and other soft parts may be preserved. Dinosaur mummies of this type have been discovered; in these instances the organic structures may have vanished completely, so that only the impressions remain. Even such fragile organisms as jelly-fish may leave a record in the form of an impression. Moulds of hard parts may also remain after the organic structure itself has been dissolved.

Soft parts may also be preserved when immersed in brine or oil, both of which may occur in some geological deposits. Preservation of this kind is more common in comparatively recent strata, for instance the celebrated Ice Age rhinoceros of Starunia in Galicia. Frozen carcasses in the permanently frozen ground of Siberia and northern North America also show partial preservation of soft parts, but geologically speaking they can only survive for a relatively short time, because the present-day situation with large frozen areas is quite exceptional in the history of the earth.

Instead of a part of the organism itself, the fossil may be some kind of record of its presence, such as a fossilised track or burrow. Fossil trackways, for instance, are frequently found on buried mud flats. These fossils are of great interest, as they give us our only chance to see the extinct animals in action, as it were, and to study their behaviour. Careful study of a fossil trackway may reveal a surprising amount of information about its author, though definite identification is only possible where the animal has dropped dead in its tracks and become fossilised on the spot. A famous fossil of this type is the body of a horseshoe crab found at the end of the long furrow drawn by its trailing, spike-like tail, and the series of prints made by its numerous feet. Sometimes ammonites – shelled marine animals – floated gently down to the sea floor in

2 Unusual circumstances led to the preservation of this dinosaur 'mummy'. This *Anatosaurus*, of the duckbill family, lived in latest Cretaceous (Lance) times in Wyoming. After death, the carcass dried out and partly mummified; finally, the beginning rainy season buried it in rapidly accumulating sediments, where the impression of the mummified skin was preserved.

*Below* front view of the animal as found, lying on its right side, head thrown back in typical death pose; *Centre* close-up showing fossilised skin impression; *Right* foreleg showing the web of skin between the toes. The webbing of the foot accords with other evidence indicating that duckbill dinosaurs were powerful swimmers.

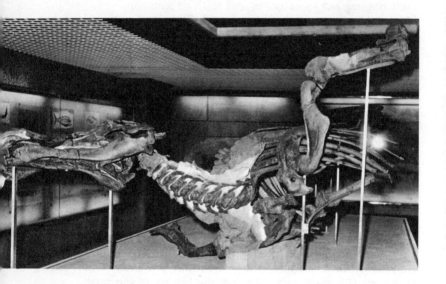

death, and left a keel-mark at the point where they touched down, before falling over on the side.

Other fossils resulting from the activity of animals of the past include tooth-marks, for instance hyena-bitten or rodent-gnawed bones, coprolites, or fossilised dung, sometimes with identifiable remains of the food taken; and eggs. Stomach contents may also be preserved; instances range from the celebrated fish-within-a-fish (figure 59) to the plant pollen found in the stomach of the frozen Beresovka mammoth.

Such are the varied records of the life of the past that palaeontologists set out to study. Some types of animals and plants, to be sure, have left no fossil record; so that the total

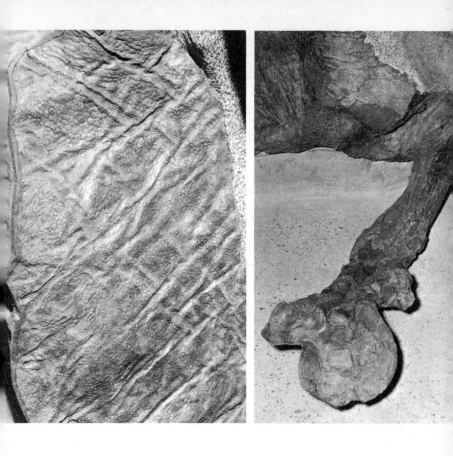

variety of fossils from any given geological epoch is inferior to that of the living world of the present day, the domain of the neo-zoologist and neo-botanist. However, it should be remembered that the palaeontologist has to reckon with innumerable successions of such living worlds, each unlike its predecessor and successor.

Fossils found in one and the same stratum may represent animals that lived at one time and place, but this is not always the case. It may happen, for instance, that fossil bones are eroded out of older strata and then become re-deposited in new strata together with new bones; a kind of *danse macabre* resulting in a temporal mixture, or heterochronic fauna. A different kind of heterogeneity results from post-mortem transport of remains, for instance by rapid streams that introduce the bones and shells of freshwater and land animals into offshore deposits alongside the remains of the local marine fauna.

This makes it necessary to distinguish between the biocoenosis, or assemblage of living beings that once existed together, and the thanatocoenosis, or assemblage of dead beings brought together by various agencies. Even the death assemblage as such does not come down to us unaltered, for the processes of deposition and the following (or diagenetic) history of the strata may wipe out the record of some of the forms present in the original thanatocoenosis. However, careful study may unravel much of this history and make it possible partially to reconstruct the biocoenosis – in other words, to make the past come alive again.

Indeed, that is one of the major objectives of palaeontology, or more precisely of the branch called palaeoecology. Interwoven with this are other branches of palaeontology, for instance the study of the process of evolution, and the stratigraphic palaeontology that attempts to record the history of entire faunas and floras in relation to that of the earth. Palaeontology, it may be seen, is very much more than simply digging up a fossil, giving it a name, and putting it into a show-case.

## Discovery of the fossil record

Fossils in the sense explained above have been known for a long time – there are even examples of fossil shells collected by the cave-men of the Ice Age in Europe. A systematic study of fossil remains may perhaps be said to begin with Leonardo da Vinci (1452-1519), who correctly regarded the fossils as remains of organisms of the past. However, the *vis plastica* theory of Avicenna (980-1037) prevailed well into the eighteenth century. Fossils were thought to be produced by a 'fertility wind' or creative force in nature, called vis plastica, which brought forth the likeness of organic things out of dead matter. At this time it was generally believed that living beings, especially vermin, could appear directly

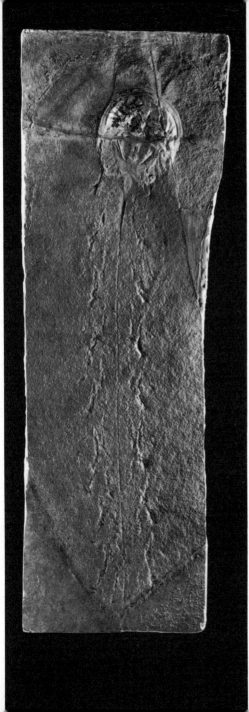

3 Fossilised horseshoe crab and trail, from the lithographic stone of Solnhofen in Bavaria (late Jurassic). Fossil trackways are not uncommon, but it is extremely rare to find the remains of the animal itself as well. This remarkable specimen shows both the crab, the track of its walking feet, and the trail of its spike-like tail. The Solnhofen genus *Mesolimulus* is closely related to the living *Xiphosura* ('Limulus') now found on the east coast of North America. The horseshoe crabs have indeed changed very little since the Palaeozoic, and thus form a famous example of a 'living fossil' or arrested evolution. In the earlier Palaeozoic, however, the horseshoe crabs evolved at a rapid rate. Evidently at that time they adapted themselves to a mode of life which they have been able to pursue ever since, so that further change would have been unnecessary or even harmful.

4 The palaeontological hoax has a long history, culminating in the 'discovery' of a supposed ape-man skull at Piltdown, Sussex, early this century. For many years this skull influenced theories on the descent of man until it finally was exposed as a composite of a human cranium and an ape jaw. The 'lying stones' of Professor Beringer are an earlier example of this sort of thing.

out of dead matter (flies out of dung, rats and mice out of carcasses), so that a kind of half-way creation would logically result in lifeless shapes imitating living beings.

It was not until the early eighteenth century, after the unfortunate Professor J. Beringer of Würzburg had, all unwittingly, published in 1726 his *Lithographiae Wirceburgensis*, that the *vis plastica* theory was finally abandoned. He had collected a series of 'fossils' which tended to become ever stranger: animals, plants, spiders' webs, the sun, planets, comets, and the name of God in Hebrew. The professor is reputed to have found, at last, his own name in the 'fossil' state, and the 'Lügensteine' or 'lying stones' were unmasked as the fabrications of two jealous colleagues. Beringer attempted to recall and destroy his book, which has become one of the curiosities of geological literature.

Meanwhile, important progress had been made in the field of stratigraphy, that branch of geology dealing with the stratified rocks and their origin, sequence, and correlation. A pioneer study was made by the Dane Nicolaus Steno (or Niels Stensen), 1638-1687, who formulated three basic laws. The first law states that, in a series of deposits as originally formed, the lowermost are the oldest and the uppermost the youngest (the law of superposition). Accordingly, the time sequence in a geological diagram always goes from the bottom up, and any other arrangement looks absurd to a geologist. In his second law, Steno states that deposits when formed have an approximately horizontal surface; so that strata now tilted at an angle to the horizontal must have been moved from their original position. Finally, Steno's third law states that any one sedimentary stratum must either cover the whole earth or be bounded laterally by other deposits, or else thin out to an edge; if sectioned layers are visible, this is due to removal of part of the stratum – for instance, through erosion.

In the early eighteenth century, Giovanni Arduino of Venice and the German J. G. Lehmann each made a threefold division of usual rocks. Arduino's classification of the sequence of strata in the province of Venice into Primary, Secondary and Tertiary units

5 The excavation of the original skull of the great sea monitor, *Mosasaurus*, from the late Cretaceous chalk at Maastricht, Netherlands. Found in 1780, it was fought over both in court and on the battlefield, finally to be removed to Paris, where it was studied by Cuvier. Figure after Saint-Fond, 1798 or 1799.

is still in part extant, for the Tertiary is still recognised.

Palaeontological thought at this time was largely dominated by belief in the biblical Deluge; fossils were thought to be remains of men and animals drowned in the Flood. The Swiss doctor J.J. Scheuchzer described some vertebrae from the Franconian town Altdorf in the belief that they were remains of a man drowned in the Flood; they have later been shown to belong to an ichthyosaur, or fish-lizard. Somewhat later he published a giant salamander skeleton, likewise interpreted as a *Homo diluvii testis* or human witness of the Deluge. This latter being, as shown many years later by Cuvier, dates from the Tertiary. It is perhaps not to be wondered at that Voltaire, carried away by anti-clerical zeal, tried to explain the fossil marine animals now found on dry land as remains of picnic meals thrown away by pilgrims.

In the later part of the eighteenth century, Jean Etienne Guettard (1715-86) devoted himself to the mapping of geological strata in France and was able to demonstrate that such strata were continuous, often for long distances, and could be characterised by their lithology and the fossils they contained; this was the birth of the concept of geological formations. In Germany, Abraham Gottlob Werner (1749-1817) studied the problem of stratal sequence and gained great support for his theory, according to which all the rocks had been precipitated out of a primordial ocean enveloping all of the earth.

With William Smith (1769-1839) the first true stratigrapher makes his appearance. Like Guettard he found that the geological formations were characterised by their fossils, but he also studied the succession of such fossil assemblages and showed that the sequence in different areas remained the same even if the lithology was different. In 1815 he published a geological map of England and Wales, giving a list of all the strata from the oldest to the youngest. Now for the first time material was on hand for a detailed chronology of the history of the earth, and students in many different countries followed Smith's example.

His contemporary Georges Cuvier (1769-1832), working out the

sequence in the Paris basin, taught that the successive periods in the history of the world were invariably closed by a great catastrophe, wiping out most of the living beings of the time, whereupon new forms of life were created or immigrated from other regions. In this way it was possible to explain why successive formations contained quite different types of fossils. The antithesis of catastrophism is the principle of uniformitarianism or actualism, according to which the geological processes of the past have been of the same nature as those of the present time; by this theory, such upheavals as volcanic eruptions, earthquakes and the like have only regional and restricted geological significance. This concept had been brought forward by the Scot James Hutton (1726-97), the German C. E. von Hoff (1771-1837) and others, and finally won the day in 1830-33 when Charles Lyell (1797-1875) published his main work, *The Principles of Geology*.

Meanwhile, palaeontological discoveries had led to new and startling glimpses of the bestiary of the past. The first came with the find of the skull of a gigantic sea reptile in 1780 in Maastricht, Holland (figure 5). The subsequent history of this specimen is remarkable. It was excavated in a subterranean quarry near the St. Peter fort, which played a part in the Napoleonic wars. The physician of the garrison, a certain Dr Hofmann, bought the skull. Now the Canon, a Dr Goddin, who owned the ground above the mine, intervened and had the skull confiscated. Hofmann appealed

6 Leaf-shaped teeth of the dinosaur *Iguanodon* are typical of herbivorous
reptiles: (a) original worn tooth found by Mrs Mantell in 1822;
(b), (c) unworn teeth from the inside and from behind. Resemblance
to living *Iguana* led Mantell to the assumption that the fossil form had
been a gigantic iguana lizard; later on it was recognised that the dinosaurs
were more closely related to crocodiles than to lizards.

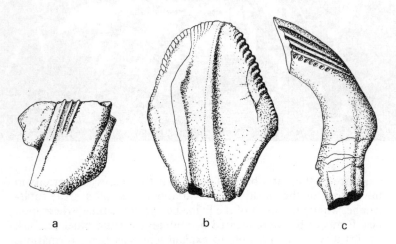

to the court, and there followed a famous case, in which Roman
law played a decisive part. Goddin won the suit and brought the
skull home in triumph, to exhibit it in his museum, a building near
the fort.

However, he did not triumph long. In 1795 the fort was attacked
by a Republican French army. The shelling was intense, but the
French studiously avoided aiming at the museum beside the fort.
Dr Goddin immediately suspected that the French planned to
secure the skull as a war-trophy, and promptly had it hidden in
the town.

When the town had been conquered by the French, a scientist
in the French army by the name of St. Fond persuaded the com-
manding officer to offer a reward of 600 bottles of wine for the
skull. Next day the skull was delivered by a dozen grenadiers,
gleefully claiming the reward.

The skull now found its way to Paris, where St. Fond published
a book about it, in the belief that it belonged to a crocodile. A few
years later it was re-studied by Georges Cuvier, who concluded that
it was more closely related to lizards and snakes than to crocodiles;

we would now say that it did not belong to the subclass Archosauria, but to the Lepidosauria. In 1828 it was given the name *Mosasaurus* ('Meuse lizard') by the British geologist W.D. Conybeare.

Cuvier's main interest lay in the fossil mammals that he excavated in the neighbourhood of Paris, and in their living relatives. He may be regarded as the true founder of the study of fossil vertebrates, and as such is indeed a towering figure in the annals of palaeontology. He was not, however, the first to identify a fossil dinosaur. This honour fell to his British contemporary, the geologist and doctor Gideon Mantell.

In 1822 Mrs Mantell, while waiting for her husband, who was visiting a patient, found a small fragment of a tooth at the edge of the road (figure 6). Mantell, unable to identify the specimen, consulted Charles Lyell, who sent the tooth on to Cuvier. In due time the answer came from Paris; Cuvier thought the tooth might have belonged to an extinct rhinoceros.

This did not seem right to Mantell, who searched for more fossils and discovered some bones and teeth, which were also sent to Cuvier. Again the problem defeated Cuvier, who was now inclined to ascribe the material to some kind of hippopotamus. Mantell himself was unable to match the teeth in the skull collection of the Royal College of Surgeons, until the curator showed him the newly acquired skull of a lizard called *Iguana*. He then realised that the fossils were those of a giant reptile, which was later named *Iguanodon* (iguana-toothed) by Conybeare. Later finds of complete skeletons have given us a detailed knowledge of this kind of dinosaur.

The first to make systematic collections of fossil reptiles was Mary Anning from Lyme Regis. She earned a living by collecting and selling fossils from the nearby cliffs. At the age of twelve she found her first fossil reptile, a skeleton of the fish-lizard *Ichthyosaurus*. Gradually she became a highly competent anatomist as well as a skilful collector, and many fine skeletons were excavated and later prepared by her. In 1821 she found the first skeleton of

7 Nicolaus Steno　　　　　　　　Georges Cuvier

*Plesiosaurus* the swan-lizard, and in 1828 she discovered the first skeleton of a flying lizard to be found in England.

In Germany, Quenstedt was a great student of the Mesozoic system, like his French colleague D'Orbigny, while Gaudry continued Cuvier's work in the Tertiary. In England, fossil vertebrates were the domain of the great anatomist Richard Owen, who coined the name Dinosauria. In America, Joseph Leidy was the founder of vertebrate palaeontology, but above all it was O.C. Marsh and E.D. Cope who in the quarter century 1870-95 opened the way to the immense fossil fields of the American West. They transformed palaeontology to a dynamic science and charged it with a spirit of discovery. At the same time, the rivalry and enmity between these two eminent scientists is a dark chapter in the history of palaeontology. Even without the rivalry, expeditions of that time encountered great difficulties. The enormous distances to be covered in pathless terrain, the gigantically big and heavy fossils that had to be transported with great care, and also the menace from the Indians, combined to make each expedition an adventure as well as a very large undertaking.

Many dinosaur skeletons are mounted in museums, but a mount of this kind is very expensive, so that only a part of the collected material is commonly exhibited in this way. In Europe, museums

William Smith                    Charles Lyell

with large collections of dinosaurs may be found, for instance in London, Brussels, Paris, Berlin, Frankfurt, Tübingen, and other cities. Both in Canada and the United States there are numerous museums with excellent exhibits of this kind.

## The discovery of evolution

Looking back at the early palaeontologists, it may seem odd that they were not evolutionists and that some of them indeed, for instance Cuvier and Owen, were active anti-evolutionists. However, the fossil evidence available to the pioneers was inadequate to reveal the presence of evolutionary lines, just as a few haphazard pieces of a great jigsaw puzzle can give no idea of its grand pattern. Cuvier, for instance, studied the succession in the Paris basin, where the alternation between continental and marine deposits fragments the record of every line of evolution. In this situation, evidence for evolution would drown in the mass of other detail except to a person actively looking for it. On this basis, Cuvier's attempt at explanation (the theory of catastrophism) was entirely reasonable.

In fact the fixity of species was practically part and parcel of religious dogma. The great Swedish naturalist Carl von Linné

(Carolus Linnaeus, 1707-78) had taught that the species remained immutably the same as when originally created by God. As the founder of modern biological systematics, his authority was undisputed. Yet, in his later work, Linné modified his opinion and arrived at an evolutionary concept; for instance, he thought that the several species of a genus could have evolved from a single ancestral form. But this seems to have been almost entirely overlooked by his contemporaries, and even now most writers regard Linné's teaching as completely anti-evolutionary.

Yet many others besides Linné soon observed that the facts of nature were inescapably pointing to an evolutionary solution. In France, Buffon (1707-88) and Lamarck (1744-1829) published evolutionary treatises; in Germany, Oken (1779-1851) and Goethe (1749-1832) joined the movement; in England, Erasmus Darwin (1731-1802) produced an evolutionist poem. However, although these students (Goethe being a skilful dilettante anatomist) produced many interesting observations and speculations, no factual evidence and no acceptable evolutionary mechanism were presented.

This is also true for the most detailed discourse on evolution, that of Jean Baptiste de Lamarck, which seems to have been utterly misrepresented by the 'neo-Lamarckists' one century later. In his *Philosophie zoologique* (1809) Lamarck envisaged a 'progression of life' from the simplest and lowliest organism to the highest and most complex. However, when reversing his scale and looking back from the level of the highest organisms, Lamarck speaks of a *degradation*, as if evolution moved in both directions. The work is highly metaphysical and must be regarded as a great vision, a culmination of the speculative, artistic treatment of evolution then in vogue. (Lamarck's strong artistic streak is also revealed by his genius as a classifier.) As to what later was to become the core of 'neo-Lamarckism', the belief that acquired characters could be inherited, this was simply a generally accepted belief in Lamarck's day, and was expressly rejected by Lamarck himself as a factor of any significance in evolution.

It is only natural that a great empiricist like Cuvier would find the romantic evolutionism of his contemporaries quite unacceptable, and that Darwin was later to label Lamarck's opinions as partly 'nonsense'.

It remained for Charles Darwin, Erasmus's grandson (1809-82). to make evolution scientifically acceptable. His contribution was twofold. In the first place, he amassed an enormous amount of evidence showing that evolution had in fact occurred. This was evidence of a calibre quite different from the speculations of the early evolutionists; it was a tremendous array of hard facts, convincing to any unprejudiced reader. Secondly, he proposed a mechanism, natural selection, to explain the evolutionary process. The theory of natural selection has also been called Darwinism. (Some people accepted the idea of evolution while not accepting natural selection as the probable mechanism.)

Various attempts have been made to dethrone Darwin and substitute some earlier student as the true pioneer; they mostly have an undertone of patriotic zeal and make a somewhat pathetic impression on the unbiased reader. As a matter of fact, Darwin's main intellectual debt seems clearly to lie with Charles Lyell, a fact duly acknowledged by Darwin himself. Lyell was not an evolutionist but his uniformitarian geology made a lasting impression on Darwin's mind. Darwinism, it may almost be said, is an application of uniformitarianism to the history of life.

The publication of *The Origin of Species* in 1859 revitalised the whole field of biological sciences. The impetus given to palaeontology resulted in important discoveries. In the first place it was realised that the observed pattern of the history of life was in harmony with the concept of evolution. The earliest life-forms known were comparatively lowly, and only able to exist in water. It was only at later stages that the terrestrial habitat, and finally the air, were gradually conquered by the organic world, and at the same time living beings tended to become ever more complex and varied.

Individual examples of evolutionary sequences were also

discovered. The Russian palaeontologist W. Kowalewsky (1843-83) brought forth the first version of the famous horse series, which was shortly afterward corrected and augmented by his American colleague Marsh. A great number of well-documented cases of this type is now known, so that palaeontology has demonstrated that evolution is not a theory but – as W. D. Matthew has put it – a fact of record.

While evolution was soon accepted as undeniable, the Darwinian explanation came in for criticism. The main criticism was succinctly formulated by Cope, to the effect that Darwinism taught of the survival of the fittest without explaining the *origin* of the fittest – in other words, the origin of the individual variations providing the material for selection. This was indeed the main stumbling block, and it caused Darwin himself reluctantly to admit the inheritance of acquired characters as part of the mechanism.

Around the turn of the century, most evolutionists were anti-Darwinian, and espoused neo-Lamarckist or teleological theories. However, both experiment and theoretical considerations indicated that neo-Lamarckism was incorrect, while the various metaphysical explanations were unacceptable to the scientists. As a result of this impasse, biologists in the early twentieth century tended to ignore the problems of evolution. The palaeontologists, however, had perforce to take an interest in these problems, and here all camps were represented, Darwinist, Lamarckist, teleologist, and finally those students who simply described evolutionary sequences but avoided discussing their causes.

The way out could be opened only by a proper understanding of inheritance. The modern science of genetics arose about 1900, although its founder, the Austrian Gregor Mendel (1822-84) had published his epoch-making work *Versuche über Pflanzen-Hybriden* as early as 1866; it remained unnoticed and was re-discovered more than thirty years later.

In brief, what Mendel showed was that inheritance was particulate, not blending as had been assumed. The units of inheritance did not blend and vanish; they remained intact for generation after generation. Though hybrids of the first generation were uniform, their offspring showed the segregation of the original parental characters in certain numerical proportions, the Mendelian ratios.

Cytological studies have since shown that the units of inheritance, called genes, are situated in the nuclei of the sex-cells (gametes), where they form a number of filamentous structures called chromosomes. Each gene has its own fixed locus in the chromosome, and each chromosome is present in two versions, one of which was received from the mother and the other from the father. Thus each gene is present in double dose. The main exception is found in the sex chromosomes, for one sex (the male sex in mammals) in heterogametic, the paternal and maternal chromosomes being unequal (or one may be missing).

Research into the microcosm of the cell nucleus is now reaching a stage when the structure of the genes themselves is being mapped. The fact that this structure may change and thus cause a change in the characters of the organism has however been known for a long time; this is called a gene mutation (chromosome mutations also occur, for instance affecting the serial arrangement of genes in a chromosome). That mutations occur was first shown by the Dutch botanist Hugo de Vries (1848-1935), one of the re-discoverers of Mendel's work. Since then the nature, effects and evolutionary significance of mutations has been studied by generations of scientists.

In the last years before World War II it was evident that evolution was regaining its place in scientific circles, but the real swing came only after the war. At this time, the evidence from various different branches of biology – palaeontology, genetics, systematics, ecology, embryology, and so on – was pooled. The outcome is called a synthetic theory of evolution, since it attempts to integrate the evidence from all of the specialised fields. It shows that evolution does, in fact, result from the interplay between mutation and selection.

Every individual in nature is a store of an enormous number of mutations. That all men, except identical twins, are different from each other is due to differences in their genomes (gene sets), and these differences have originated by mutation. Further, the store of mutations in a living population is constantly being replenished by new mutations. Thus, modern genetics has answered Cope's question about the origin of the variations in nature.

Natural selection, however, does not act directly on the mutations as such, but on the populations of individual organisms. A constant reshuffling of the genes is being effected as a result of the sexual reproductive system. Not only are the sets of chromosomes shuffled at the formation of each gamete, but in one phase of the process maternal and paternal chromosomes even exchange parts ('crossing over'). All this leads to a never-ending recombination of genetic material. The experiments carried out by Nature itself show that this system is indeed the only one conducive to progressive evolution, for all those organisms that reproduce asexually tend to remain at a lowly, primitive stage.

Selection is now seen as the agency which affects a systematic change of gene frequencies from generation to generation. The effect is a statistical one, and even a very slight advantage will inexorably affect the outcome in the long run. Thus, for instance, specialisations may be driven to a precision vastly exceeding what might seem reasonable to us. The actual mechanism through which selection operates may consist in differential fertility, differential length of life, or differences in the capacity to find a mate. Usually the result is such as to increase the adaptation of the organism to its current environment, though some types of selection may have the opposite effect (the selection for beautiful display organs in some birds makes them conspicuous and so easier to catch).

While mutations are practically random – in the sense that they have no relationship to the current needs of the organism – selection is a strongly orienting factor, and continued selection in the same direction for a very long time will bring about an evolutionary trend. The discovery of such trends has become a commonplace in palaeontology; many examples are noted in the following pages.

The species is the basic evolution unit, for only within species is gene exchange and re-combination possible. Once two distinct species have formed, an irreversible evolutionary step has been taken; while infra-specific changes may, in principle, be reversible.

## Geological time

The diluvialists taught (and a few cranks still do) that *all* the extinct animals existed together before the Flood, and were killed off by this catastrophe. However, the work of stratigraphers revealed that a long succession of different faunas and floras must be postulated, and the uniformitarian geology made it necessary to accept a very long chronology for the earth, perhaps to be measured in hundreds of millions of years. This was something quite new in Western thought. Calculations of the age of the earth had previously been made on the basis of the Scripture genealogies, and the most generous figure obtained in this way was a mere 6,984 BC for the creation of the earth. Other world calendars based on myth and tradition tended to be slightly longer; the ancient Persians thought that man had existed for 12,000 years, the Egyptian figure is approximately the same, and Plato believed that Atlantis had been destroyed some 9,000 years before his time.

Some Oriental calendars tended to greater length. The Chaldees of Mesopotamia estimated the antiquity of man at 473,000 years, but that of the earth at more than two million years. Their mythology is exceptional in conceiving of such a long span of time for the earth before the coming of man. Chinese calendars also were of this order of magnitude, while the Hindu reckoning outstripped all the others by operating with world ages of no less than 4.32 billion years. Such a period, corresponding to one day in the life of Brahma, would equal the total existence of the earth, but to date only somewhat less than one-half of the world age was thought to have elapsed. By a remarkable coincidence the length of the Hindu World Age is very close to modern estimates of the present age of the earth.

8 Excavating a skeleton of the giant sauropod *Brontosaurus* in the late Jurassic Morrison beds of North America.

These speculations, however, were quite unknown to Western geologists and palaeontologists, whose outlook was necessarily coloured by the short chronology. With the apparent stability of living species in view, who could reasonably maintain that the incredibly diverse living beings could have arisen by evolution within a few paltry thousands of years? The new concept of geological time presented just before Darwin entered the field is clearly a crucial factor in the development of evolutionary thought. In this new scheme there was plenty of time for the gradual evolution of higher forms out of the lowliest ancestor imaginable, and perhaps even out of inorganic matter.

Lyell calculated the length of geological time on the basis of the changes in the marine mollusc faunas during the Tertiary and Quaternary periods. Using the figure of 1 million years for the Quaternary as a guide (this was, of course, no more than an informed guess) he arrived at the estimate that the Tertiary period began about 80 million years ago. A glance at the modern chronology on page 35 will show how excellent a result this was. On the other hand, his estimate for the time since the Cambrian was only 240 million years, little more than a third of the measurement given here. Still, Lyell's figures were certainly of the right order of magnitude. Lyell's results remained among the best until Reede in 1879 estimated the span since the Cambrian at 600 million years, on the basis of sediment thickness and estimates of the rate of sedimentation.

At about this time, however, a setback was introduced by the famous calculation of the age of the earth published by Lord Kelvin in 1883. He proceeded on the assumption that the earth had formed as a molten sphere that had then gradually cooled off. It could then be shown, if you started with a globe with an even temperature of 4,500° Centigrade and ended up with one in the present condition of the earth, that the surface of our planet would have become solidified at some time between 20 and 40 million years ago. Hence the entire geological history of the earth would have to be compressed within that time.

This calculation made Kelvin a real thorn in the flesh for geologists, whose claims regarding the age of the earth appeared to him grotesque and presumptuous. The geologists, on the other hand, thought it impossible to force the immensely complicated geological history, and the evolution of life, into such a comparatively brief stretch of time, but were unable to find the fault in the calculation.

The error lay in the omission of radioactivity, which constantly produces heat, which means that the earth is not in reality cooling. Ernest Rutherford, one of the pioneers in the study of radioactivity, was aware of its significance for the age of the earth at an early stage, and also realised that radioactive minerals might be used to estimate geological time. These ideas were soon developed further by other students. The principle is to analyse the amount of a given radioactive substance in a rock, as well as the amount of its radiogenic derivatives; if the rate of decomposition is known, a direct calculation of the age of the rock is possible.

The first radioactive minerals to be used in this way were compounds of uranium and thorium, both of which produce isotopes of lead as end-products (helium is also produced). The main drawback was that both elements are fairly uncommon in nature, so that measurements were only possible for a few types of rock. Use of sedimentary rocks was almost out of the question, and so it was very difficult to tie in the uranium and thorium dates with the stratigraphic succession. Some dates, however, were secured in this way, and the order of magnitude of the chronology could be determined.

In recent years these methods have been largely superseded by the potassium-argon method of age determination, which has been perfected by J. F. Evernden and his co-workers at Berkeley, California. Potassium contains, in small but constant amounts, an isotope with the atomic weight 40, which is radioactive. It will disintegrate into calcium or argon with the same atomic weight. The task is now to measure the amounts of potassium and radiogenic argon in the rock sample to be dated. There are various difficulties, as argon is a gas also found in the atmosphere. The problem, however, has been mastered, and the method has turned out to be very reliable, at least insofar as it gives highly consistent results. Potassium is one of the more common substances in Nature, which is of course highly desirable. Ideal material for this kind of study is furnished by volcanic tuffs, which have been incorporated into sediments. At the time of the eruption all the argon formed up to then would have been lost, so that the radioactive chronometer would be set back to zero.

Measurements now suggest that the age of the earth is between 4.5 and 5 billion years – more than a hundred times the value suggested by Kelvin. The time since the beginning of the Cambrian (about 600 million years) is little more than one-tenth of the total age of the earth. For comparison it may be mentioned that the first urban cultures arose some 5,000 years ago, or a millionth of the age of the earth. Thus insignificant appears the scale of human history in comparison with that of geology.

Table 1 gives an outline of the chronology of earth history and a summary of the main forms of life of the various periods. Up to the time of radiometric dating this was a relative chronology only,

**Table 1** Earth history and chronology

| Era | Period | Age in millions of years | Dominant forms of life |
|---|---|---|---|
| Cenozoic | Quaternary | 0–3 | Man, ape-men |
|  | Tertiary | 3–63 | Mammals, birds |
| Mesozoic | Cretaceous | 63–140 | Dinosaurs and |
|  | Jurassic | 140–180 | other reptiles |
|  | Trias | 180–230 | Reptiles |
| Palaeozoic | Permian | 230–280 | Reptiles |
|  | Carboniferous | 280–345 | Amphibians |
|  | Devonian | 345–405 | Fishes |
|  | Silurian | 405–430 | Arthropods |
|  | Ordovician | 430–500 | and other |
|  | Cambrian | 500–600 | Invertebrates |
| Precambrian times |  | 600–c. 5,000 | Incompletely known |

that is to say, without any dates. For instance, one knew that the Jurassic preceded the Cretaceous, but as to the amount of time involved this was anybody's guess.

That did not really bother the early stratigraphers, whose main interest lay in arranging the geological strata in their correct temporal sequence. While geological time is divided into periods, the corresponding rocks form so-called systems of strata, and the recognition and naming of these systems was the first great achievement in historical geology. This work was carried out in the first half of the nineteenth century, following William Smith's work, and was brought to a conclusion with the naming of the Ordovician in 1879. Much of the work was done in England and Wales, so that it is instructive to refer here to a geological map of this important type-area in geology (figure 9).

If we imagine the Recent and Ice Age deposits (often called 'alluvial' and 'diluvial' respectively, especially in older literature) stripped away, the youngest deposits are those in the London and Hampshire basins and East Anglia. These rocks form the Tertiary System, studied by Lyell.

North and south of the London basin, however, we get into older strata. The reason for this is that the London area forms a gentle downfold or syncline, the axis of which approximately coincides with the course of the Thames. This syncline is of long standing, for the deposits accumulated in it begin with the early Tertiary London Clay. To the south lies the Wealden Anticline, a raised fold parallel to the syncline north of it. The summit of the anticline has been cut across by erosion, so that gradually older strata are exposed as the axis of the fold is approached. The rocks that are exposed here underlie the Tertiary ones and comprise the Chalk, the Greensand, and the Weald Clays, all of which are grouped together in the Cretaceous System (derived from the Latin *creta*, 'chalk').

Proceeding to the northwest from London, we again get first into Chalk (Cretaceous) country, but then cross successively older deposits. This is because the whole sequence of strata is slightly canted, dipping down to the south-east, so that strata hidden far below the surface in the London area gradually approach the surface and become visible to the northwest.

We now arrive at a belt consisting of the strata termed Oxford Clay, Oolites, and Lias, and which belong in the Jurassic System. As the name implies, the type area is not here but in the Swiss Jura mountains, where the system was named as early as 1795 by Brongniart and Humboldt. Having traversed the Jurassic, and approaching Birmingham, we reach the great area covered by New Red Sandstone. This thick unfossiliferous sandstone formation yielded little information and was described as underlying the Jurassic and overlying the Carboniferous. Evidence on this part of the geological column had to come from other areas.

In Germany, however, a sequence of well-characterised deposits

9 Geological map of the British Isles. Unusual complexity and completeness of the geological sequence combine to make this one of the most important stratigraphic type-areas of the world. Much of the pioneer work establishing the chronology of the earth was done in England and Wales in the early nineteenth century.

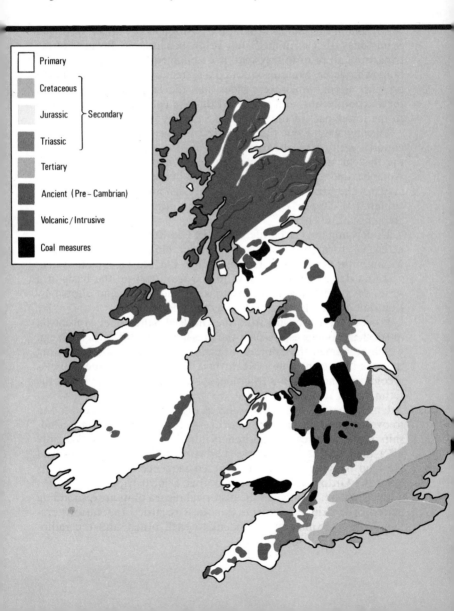

was found in the same stratigraphic position as the upper New Red Sandstone – that is, immediately below the Jurassic. Because of the tripartite nature of this system, it was named the Trias by Alberti. Somewhat later, Murchison found a system of rocks in the Russian province Perm, which were older than the Trias but younger than the Carboniferous, and so named this the Permian. It corresponds to the lower part of the New Red Sandstone.

Turning west from the New Red, we arrive in Carboniferous country with its Coal Measures. Again the type area is here, and the system was named by Conybeare and Phillips. In North America the Carboniferous is usually divided into two separate systems, the Mississippian ( = Lower Carboniferous) and Pennsylvanian ( = Upper Carboniferous), but they are not generally recognised elsewhere.

Underlying the Carboniferous is the Old Red Sandstone, another extensive formation covering considerable areas in Hereford, Monmouth, etc. It is of Devonian age, but the Devonian system was first described in Devonshire and Cornwall on the basis of a series of graywackes and limestones found there, and shown by Lonsdale to be of the same age as the Old Red.

Proceeding downward from the Old Red Sandstone, Murchison in southern Wales established the presence of a still earlier system, which he named the Silurian, after the name of a tribe inhabiting the area in ancient times. Meanwhile, Sedgwick in northern Wales recognised and named a still older system, the Cambrian (from the Latin name for Wales).

Unfortunately, the two systems as originally defined were found to overlap, and the ensuing controversy turned the previous friendship of the two great stratigraphers into lifelong hatred. It was only after their deaths that Charles Lapworth in 1879, finally resolved the feud by singling out the disputed sequence as a system of its own, the Ordovician (also named after a local tribe).

Each system, then, gives the historical record of its corresponding period. The original students doubtless regarded the various systems as approximately equivalent to each other, and the radiometric dates in fact show that the periods tend to be of about the same length (with the exception of the very short Quaternary period), although some variation does indeed occur; for instance, the Silurian is a comparatively short period, whereas the Cretaceous is certainly one of the longest. On an average the periods lasted some 50-60 million years.

As shown in the diagram, periods are grouped together into longer divisions called eras: the Palaeozoic, Mesozoic, and Cenozoic. The eras lasted about 400, 170, and 65 million years respectively. On the other hand, periods and systems are also subdivided, though this is not shown in the present diagram. A period is divided into epochs, and the corresponding part of a system is termed a series. Again, epochs and series are divided into ages and stages respectively. The average length of an epoch is some 15 or 20 million years, and that of an age about 5 million, but individual values vary greatly.

## Palaeogeography and continental drift

Geography deals with the face of the earth: its seas, lands, climates, and its human populations. To us, who observe them only during the short span of a human life, the continents and seas, the mountains and great rivers seem eternal, changeless. But in geological perspective they are all seen to be in a state of flux; even the solid rock under our feet may be ephemeral. The geography of the earth has changed greatly in geological time.

To make a palaeogeographic map, showing the face of the earth at a given moment in geological time, you start out by marking down all the known deposits dating from that moment. Their distribution will give the basic facts. Beach deposits indicate the position of the ancient coastline; marine and continental strata indicate the primary division between land and sea. Gradually, a picture of the contours of the land will emerge, though of course in many areas deposits from the period are missing, so that the situation in those areas cannot be directly observed.

10 Gondwanaland, the southern supercontinent, reassembled at the point of initial fragmentation in mid-Mesozoic times. Circle shows approximate position of South Pole, dots denoting individual determinations; sample areas indicated by circles and crosses. Palaeolatitudes as numbered. After Irving.

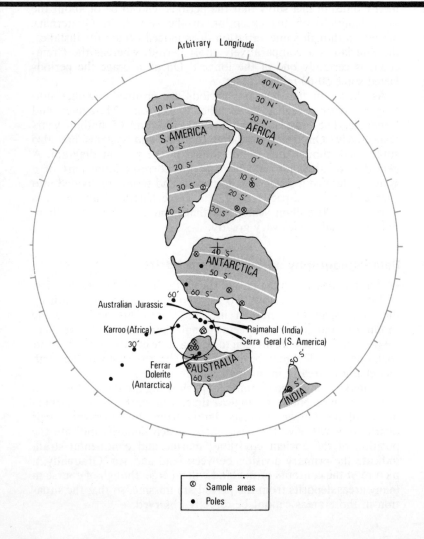

It is often possible to say something about such blank areas too. For instance, the presence of great masses of delta sediments indicate that the coastline we have traced belongs to a large land mass, sufficient to feed great rivers. Other information on the unknown land may be found in the nature of the sediments (which may show whether they were derived from high mountains or from plains, for instance) and from the fossil remains of land animals and plants carried out by the streams. These may also give a clue to the climate. Of course, more direct evidence of conditions on land is found in the continental deposits; rivers, lakes, marshes, deserts, all produce strata that tell the story of their origin.

Conditions in the ancient seas are even better known, especially for the older systems, which are now mainly represented by marine deposits. The sediments formed in shallow and deep waters can usually be told apart, and their fossils will again give information on the conditions of life. Corals, as we know, live only in shallow, clear, warm waters; thus the presence of fossil corals indicates both the depth of the water and the climate. In this way we know the depth zones of the seas of the past, in the areas where they encroached upon the present-day lands.

Fossil assemblages from different parts of the world will give important information on the connexions between continents, or between seas. Great similarity between two faunas indicates that there was an open migration route between them – a land bridge for two land faunas, a marine strait for two sea faunas. Conversely, marked differences between two contemporary faunas will indicate the presence of barriers. If the two dissimilar faunas are close to each other geographically, the barrier was almost certainly a physical one (for instance, an isthmus separating two bodies of water); in other cases such differences may, for instance, reflect climatic zonation.

The sea level greatly influences the interchange of local faunas. A rising sea level (this process is termed a transgression) will tend to break off migration routes for land animals, and open new corridors for marine animals. For instance, melting of the Antarctic ice cap would result in a global rise in sea level of the order of 300 feet. This would flood the lower coastal plains completely, cut off the land bridge between Africa and Asia, and open a migration route for marine faunas between the Mediterranean and the Indian Ocean. On the other hand, a similar regression, or lowering of sea level, would lead to emergence of a land bridge from northeastern Siberia to Alaska, and at the same time cut off the Arctic Sea from the Pacific. In the fossil record, a sudden influx of new forms may often be interpreted as the result of the opening of a new migration route.

One feature noted by the old authors continued to intrude itself upon the minds of geologists – the presence of marine fossils in high mountains. In time this was recognised to be not a freakish exception but in fact a general rule. It finally led to a new theory of the birth of mountain ranges. They were found to originate as so-called geosynclines, great elongate troughs that subsided gradually for many million years, and in which there would accumulate a long succession of deposits derived from flanking land areas. At a later stage, the geosynclinal strata became folded and uplifted to form a chain of mountains. For instance, the Alps arose in the Tertiary period out of a great geosynclinal sea called Tethys, which had been gradually forming and deepening ever since the Trias. The folding, of course, implies considerable lateral movements of the coastal blocks on one or both sides of the geosyncline; in the case of the Alps it has been estimated that the original geosyncline was at least ten times as broad across as the existing mountain range.

The geosyncline concept was formulated by the American geologist J.D.Dana, developing ideas previously voiced by his compatriots J.Hall and C.E.Dutton. It would seem that the continents have in fact been built up by successive geosyncline cycles of this kind, for old continental blocks like the Baltic or Canadian shields consist of the roots of very ancient mountain ranges that have long ago been levelled by erosion. The orogenic or mountain-building phases are separated by long periods of comparative quiescence. For example, the Alpine orogeny, which is the most recent, was preceded in Europe by the Variscan (Permian) and Caledonian (Devonian) orogenies, plus a number of Precambrian pulses. It would also appear that times of orogeny bring about a cooling of climates and a greater differentiation of climatic zones.

The great lateral movements evinced by the intricate folding and overthrusting suggested logically the phenomenon that has been termed continental drift. Lateral movements may indeed be observed at the present day. In California the great San Andreas fault extends from San Francisco to the southeast, and forms the boundary between two blocks that slide past each other in jerks. The 1906 earthquake represents one such jerk, in which the lateral displacement was of the order of three metres.

Dana showed that the main parts of the earth's crust were balanced as if floating on a denser substratum; according to this, the principle of isostasy, the continents were in fact buoyed up like icebergs by the deeper, heavier material forming the mantle of the earth (which, in turn, envelops the still heavier core). The boundary between the lighter, floating crust and the underlying mantle has been found to affect the seismic waves caused by earthquakes; it is thus a prominent feature in the structure of the earth, and has been named after its discoverer, the Croatian scientist A. Mohorovičić. The physical nature of the Mohorovičić discontinuity (or 'Moho' for short) may be disclosed in future, if and when it becomes possible to make a deep boring to this level (a 'Mohole'). Meanwhile, the principle of isostatic floating suggests the possibility of quite extensive movements.

It had been noticed as early as the middle of the nineteenth century that the eastern coast line of America resembled the west coast of Europe and Africa, and there was some speculation as to whether these continents might once have been united. It remained, however, for the German scientist Alfred Wegener to investigate this problem systematically. His theory of continental drift was published in 1915 in a book entitled *Origin of the continents and oceans*. He assembled an impressive body of evidence suggesting

11 The San Andreas and its subsidiary faults indicate the boundary between two great blocks, of which that on the Pacific side is moving north-westwards in relation to the continent; this movement has gradually opened up the Gulf of California. In aerial view, the San Andreas stands out dramatically as a demarcation line between two different landscapes, as in this view along the fault, looking north-west across Indio Hills, Riverside County, California.

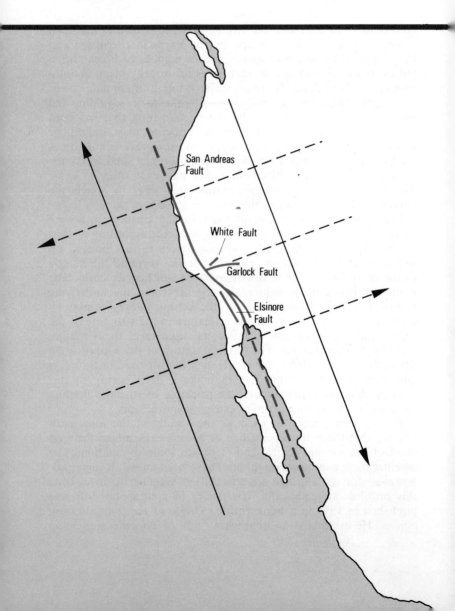

that the southern continents once formed a single supercontinent, the Gondwanaland, and that they had drifted apart at a comparatively late date in geological history. The evidence includes the direct continuation of numerous geological features from one continent to another – features of various ages up to traces of the Permo-Carboniferous Ice Age which seems to have occurred simultaneously in all the southern continents.

In the half-century that has passed since Wegener made his theory known, it has been variously supported and attacked. For a long time the difficulties seemed insuperable. Since about 1950, however, new methods of study have revolutionised our views on this problem, and most students have been convinced by the evidence in favour of continental drift, although its timetable is now somewhat different from that advanced by Wegener.

The most important line of research is the study of palaeomagnetism, introduced by P. M. Blackett and K. Runcorn. This consists in determination of the magnetic orientation of the charged particles in some rocks (the principal minerals used are magnetite in volcanic rocks and hematite in sedimentary rocks). The magnetic field of these rocks is parallel to the magnetic field of the earth as it existed at the time of formation of the rock, which makes it possible to calculate the position of the magnetic poles at the time in question. A great number of studies of this kind has been made, and as a result we can now prepare a fairly accurate travel log for the North Pole during geological time.

If we determine the position of the North Pole on the basis of European rocks of different ages, we find it well out in the Pacific at the beginning of the Palaeozoic, near the present-day equator, and about half way between New Guinea and South America. It then moved in an arc, at first to the west, then gradually curving to the north, so that it was somewhere near Japan in the beginning of the Mesozoic. In the course of the Mesozoic, the pole proceeded northward, and entered the Arctic Sea basin perhaps in the early Tertiary.

On the other hand, if we determine the positions of the pole on the basis of North American rocks, a slightly different route is obtained, being transposed about 30° to the west of the one just described, except in its final part. If we imagine North America moved 30° eastward, so that its continental shelf is brought into touch with those of Europe and Africa, the two polar routes will coincide. It would then seem that America was indeed in contact with the Old World for most of geological time. This situation apparently persisted up to the Trias, when the continents started drifting apart. The drifting may still be going on; however, it appears that the phase of drift may represent a comparatively short episode in the history of the earth.

As regards the southern continents, even more dramatic changes have been indicated by the palaeomagnetic studies. It now seems clear that the supercontinent, Gondwanaland, existed almost to the end of the Palaeozoic. In the Carboniferous, the South Pole was situated in the southern part of Africa; the remainder of Gondwanaland consisted of peninsular India, Australia, Antarctica, and South America, counting clockwise from Africa. Fragmentation started in the Permian and Triassic, while the main part of the actual drift occurred in the Jurassic and Cretaceous.

These great movements seem to have occurred at the rate of a few centimetres a year, or about the same rate as the average for the San Andreas fault. The movements probably result from the continents being dragged along by currents in the underlying mantle. H.C. Urey has suggested that the mechanism is one of convection currents fed by the heat of the molten core of the earth. The phenomenon is analogous to that found in the atmosphere. The heated material from the bottom of the mantle will rise towards the Moho, then move laterally while gradually cooling, finally to start sinking at another point. The current will thus move in a circular path, forming a so-called circulation cell.

The lateral currents beneath the Moho will tend to position the continents above the zones where convection currents are descending. The lateral pressure will also result in folding and mountain-building in such zones. On the other hand, features such

12 *Above* Section of outer part of earth, showing mantle currents and crustal features. New material is added to the crust in zones of upwelling, where currents diverge. These tend to rip the crust apart, creating rifts flanked by ridges. Continents are posited over zones of convergence and downwelling, where initial downsucking creates a geosyncline, which later becomes compressed into a mountain range. *Below* Scheme showing relationship between size of molten earth core and number of convection cells in mantle.

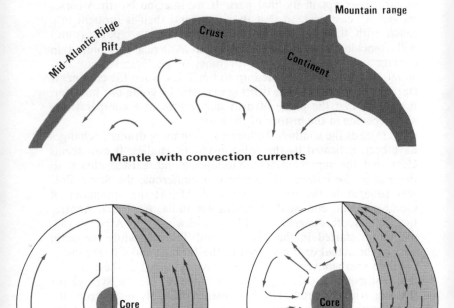

as the mid-Atlantic ridge with its internal rift would correspond to zones with upwelling mantle currents.

The number of circulation cells formed in the mantle depends on the relative size of the molten core. The larger the core, the greater the number of circulation cells in the mantle. The core now has a radius of 55 per cent of the total radius of the earth, which according to Runcorn corresponds to a differentiation of five circulation cells.

From these theoretical considerations it may be concluded that a change in the number of circulation cells, for instance from four to five, would result in a rearrangement of the continents. Runcorn suggests that such a change in fact occurred, and resulted in the observed phase of continental drift. The change is argued to be due to a gradual increase of the size of the molten core, resulting from a rise in the internal temperature of the earth. In fact, Urey in 1952 suggested that the earth did not originate as a molten sphere, but as a solid body. Heating – for instance by radioactivity – would then gradually bring the innermost core to the melting point, and then slowly increase the size of the molten mass. At the same time the number of circulation cells in the mantle would increase from one to two, then to three, four, and five, each increase corresponding to a phase of drift. Although the earlier phases are now difficult to study, there is in fact some evidence of their presence. Thus consideration of continental drift will take us step by step to speculations about the origin of the earth itself.

Meanwhile, the concept of drift is bringing new stimulus to the study of plant and animal palaeogeography, and gives a new perspective of the environment in which life evolved.

# 2 Earth history before the dinosaurs

## The Precambrian

Calculations suggest a probable age for the earth of between 4,000 and 5,000 million years. The oldest rocks known (from São Paulo in the Atlantic) have a radiometric date of 4,500 million years. Next come some rocks from Kola Peninsula, dated at 3,400 million years.

Precambrian rocks are only found in restricted areas of the world: the Canadian shield, the eastern part of South America, the Baltic shield or Scandis in Europe, the Indian peninsula, the Angara shield in Siberia, very large tracts in Africa, the western half of Australia, and Antartica. Since older rock systems tend to be covered by younger ones, the available outcrops of Precambrian rocks are comparatively limited; although they represent at least 80 per cent of geological time, they make up only about 20 per cent of the surface rocks.

The Precambrian shields usually have low relief and consist of the roots of ancient, completely eroded mountain ranges. Several different systems of such roots indicate that a series of orogenic cycles was enacted before the Cambrian; in Fennoscandia, for instance, three or four successive cycles have been distinguished. They were probably accompanied by significant climatic shifts. However, even in its most agreeable moments, this would have been an inexpressibly bleak and alien world. The continents were completely bare of life, and so lacked the ameliorating effects on climate produced by vegetation. Winds would howl across a desert land; streams would scour the hillsides and carry off all the loose material. The air was probably too poor in oxygen to keep any higher animals alive; the atmosphere has received its present store of oxygen from the metabolic activity of green plants ever since their first appearance at some unknown time in the Pre-cambrian.

At a still earlier time, life itself originated on the earth. Probably the earliest forms of life were simply big molecules that were able to reproduce by building up images in their own likeness, utilizing available sources of chemical energy. In our present-day world,

13 Remains of a very foreign-looking animal world in the late Precambrian deposits of Ediacara, Australia. (a) *Rangea* and (b) *Charnia* are frond-shaped structures somewhat resembling certain coral types. (c) *Tribrachidium* may be related to echinoderms, but is unique in having threefold radial symmetry. (d) Spriggina has been interpreted as an annelid worm, but also as a trilobite ancestor lacking outer shell. These animals lived more than six hundred million years ago.

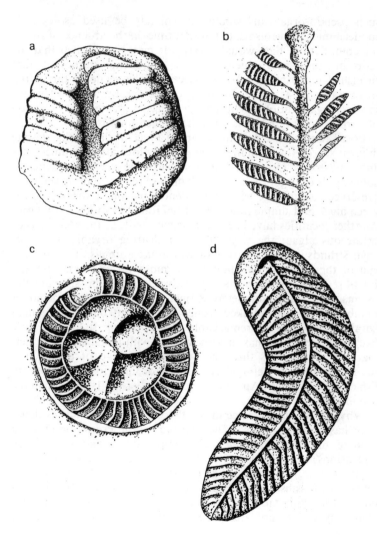

such pseudo-organisms would immediately be used as food by modern micro-organisms, but at that time, in the absence of superior competition, they would survive. It may be assumed that the reproduction was frequently faulty, producing mutations that furnished material for natural selection. The scene of this initial stage is usually assumed to have been the sea, but coastal lagoons in which evaporation might have produced a greater concentration of carbon compounds, may also have been important.

Fossil remains are extremely rare in Precambrian rocks, despite diligent search for them in the more likely formations – limestones and sandstones. Occasional finds, however, prove the existence of some kinds of organisms well back in the Precambrian. A very well preserved assemblage of clearly identifiable plant forms – blue-green algae and simple fungi – has been found in southern Ontario. At other localities have been found such fossils as the secretions of calcareous algae; the trails of larger drifting organisms (longer algal strands?); and burrows of worm-like animals. Towards the end of the Precambrian, however, a more varied fauna appears. Fossil remains include primitive bivalves, jellyfish, possible annelid worms, and some completely extinct types. The latter include a number of peculiar stalked fronds reminiscent of the gorgonian corals, as well as the remarkable *Tribrachidium*. It looks a little like a sea-urchin but has three bent arms with tentacles along their margins; it thus has threefold symmetry instead of the fivefold radial symmetry of echinoderms. Many of these fossils come from Ediacara in southern Australia, and are only slightly older than the Cambrian.

Why are fossils so scarce in the Precambrian? Various explanations have been suggested. The later part of the Precambrian was a time of mountain-building, cold temperatures, and world wide regressions; so that deposits of a type likely to contain fossils would naturally be somewhat restricted. It also seems probable that hard, fossilisable shells and tests were developed by different phyla at a comparatively late date. Even in the Lower Cambrian, many phyla that were later to develop calcareous shells, are absent or represented mainly by forms with chitinous, phosphatic, or siliceous shells, presumably more primitive. Thus the Precambrian record may be so poor because the animals of the time lacked hard parts that could be preserved.

## The Cambrian

Despite the transition indicated by some late Precambrian faunas, in most areas there is still a dramatic change from the emptiness of the Precambrian to the plenty of the Cambrian. This is reflected not only in the number of life-forms that seem to appear all at once, but also in the trails and burrows that record their activities.

In most regions the Cambrian sea transgressed (advanced) over much older rocks, and the nature of the earliest Cambrian sediments may suggest recent recovery from the infra-Cambrian ice age and orogenic disturbances. Europe in the early Palaeozoic was mostly covered by water. The Scandis, or Baltic block, and the Russian platform were partly emergent, partly covered by shallow seas. These were bordered to the northwest by the Caledonid geosyncline, extending along the axis of the present-day Scandinavian mountains into Scotland and Ireland. At right angles to this extended the mid-European geosyncline from the North Sea basin across Germany and Poland, bordered in the south by a geanticline. Most of southern Europe formed another geosyncline, while Spain was partially emergent as part of the Iberian geanticline.

North America was flanked by geosynclines along both the Pacific and Atlantic coasts of the present day, while the core of the continent was formed by the Canadian shield. In other continents, too, the ancient Precambrian shields were emergent or inundated by shallow seas, while great systems of geosynclines surrounded them.

As the Cambrian period progressed (and it lasted perhaps 100 million years), climates tended gradually to become more equable; the effects of the infra-Cambrian orogeny were wearing off. In such an untroubled world, life was unfolding in increasing variety.

The continents were still without life, except probably for some mosses covering patches of wet ground, but the seas teemed with organisms. The most conspicuous animals in the Cambrian were the trilobites, arthropods somewhat resembling wood-lice in external appearance but generally regarded as more closely related to spiders and scorpions than to the true crustaceans. The primitive trilobites typical of the earlier Cambrian had a large head shield with recurved spines extending from the cheek region, a segmented body, and a small tail shield or pygidium. The name trilobite alludes to the longitudinal lobation of the body, with an axial lobe (forming the so-called glabella on the head shield) and two side lobes (forming the cheeks of the head shield, and the so-called pleura of the body segments).

The trilobites had two well developed compound eyes, situated on top of the head shield to each side of the glabella. The mouth was on the ventral side of the head. Each body segment carried a pair of walking appendages and gills. When in danger, the trilobite would roll up somewhat in the manner of an armadillo. The pleura terminated in spines, so that the rolled-up trilobite would form an unpalatable-looking spiny ball.

Trilobites of this general type, such as the primitive Mesonacidae of the early Cambrian, and the genus *Paradoxides* of the Middle Cambrian, form important guide fossils; in fact the trilobites furnish the most important fossils for Cambrian stratigraphy.

However, even in the Cambrian more specialised trilobites may be found, with a pygidium as large as the head shield, and the number of segments reduced, to only two in extreme cases. These trilobites (*Agnostus* and related forms) are only a few millimetres in length, completely eyeless, and may be found in enormous numbers in some rocks.

Most trilobites were probably bottom-living, some were burrowing forms, and they include both mud-eaters and carnivorous types.

The mid-Cambrian Burgess shale of British Columbia is noted for the exquisite preservation of its fossils, and here are found,

14 Early Palaeozoic geography of Europe, showing the stable continental blocks separated by a system of geosynclines and geanticlines. Parts of the three blocks (the North Atlantic, the Scandinavian–Russian, and the African) were often inundated by shallow seas.

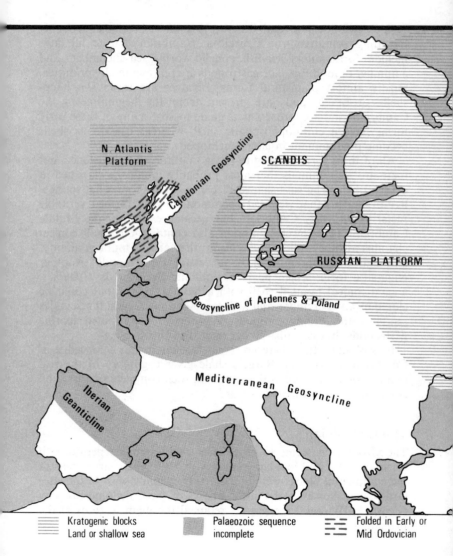

besides trilobites, several other forms of arthropods, probably including primitive true Crustacea. Cambrian marine life also includes foraminifers, jellyfish, annelid worms, sea-cucumbers, and a number of molluscs. The gastropods are represented by primitive, mostly almost planispiral forms. Bivalves are rare. The lampshells or brachiopods are present from the beginning of the Cambrian, and are important as guide fossils; many of these early forms have a chitinophosphatic shell, whereas calcareous shells become predominant in later brachiopods.

All of these forms belong to phyla that are still in existence, but there are also animals that most probably should be classified in extinct phyla. The vase-shaped Pleospongea are somewhat sponge-like organisms that were very common (in fact, formed veritable reefs) in the early Cambrian, but became extinct in the Middle Cambrian. The Lower Cambrian also contains the small conical chambered tubes called *Volborthella*; they are of doubtful affinity, although once thought to be ancestral cephalopods. Another enigmatic form, which persisted into the Permian, is *Hyolithes*, a semi-conical pointed shell with a basal aperture that could be closed by an operculum, and with a pair of appendages. It resembles remotely the living pteropods, a group of wing-footed marine gastropods, but is probably not related to them.

It is obvious that there are great lacunae in our knowledge of Cambrian faunas; the Burgess shale gives a glimpse of the real richness of animal life at this time, but such complete preservation is very unusual in the fossil record.

## The Ordovician

The Ordovician seems to have been a very equable period with warm climates and low-lying lands with little relief; large areas of the continental blocks were inundated by shallow seas. In the later Ordovician, however, the initial phases of the Caledonid orogeny occurred. The palaeogeography retained the main features of the preceding period.

15 Typical Cambrian fossils include trilobites, brachiopods, and some forms of uncertain affinities. Trilobites range from large, spiny, many-segmented forms like (a) *Paradoxides* to tiny, eyeless, two-segmented *Agnostus* (b, much enlarged). Enigmatic forms are (c) *Hyolithes* and the minute chambered shell (e) *Volborthella*, while true crustaceans may be represented by such forms as (d) *Waptia* from the Burgess shale. Among brachiopods primitive forms like (f) *Lingulella* are important; figure shows inside of valve with impressions of opening and closing muscles.

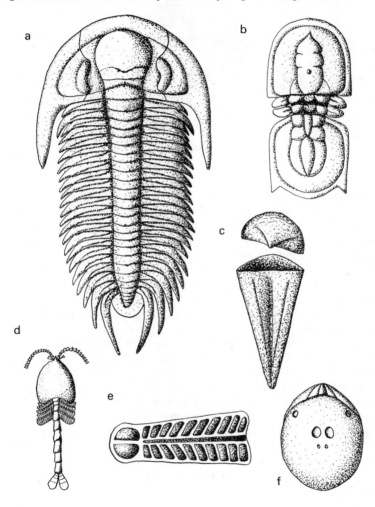

16 Floating community of several graptolite colonies of the Middle Ordovician *Diplograptus*. Each colony consists of numerous individuals arranged in rows (two rows in this form, but others may have one or four). Colonies are united into a community with bell-shaped central float and reproductive pouches.

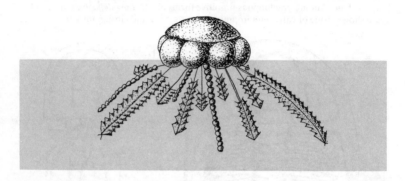

The trilobites, which now reached the height of their diversity, continue to be valuable guide fossils, especially in the limestones formed in the shallow marginal seas; but in the geosynclines a new group called Graptolithina furnishes the main chronological evidence. The graptolites were colonial animals with a chitinous skeleton. The individuals were arranged in a regular pattern along branches, and were protected by tube- or cup-shaped thecae. The primitive graptolites of the basal Ordovician formed many-branched, tree-like colonies; in later graptolites there was a reduction of the number of branches, down to a single branch in extreme cases.

The graptolites attached themselves to floating objects, or to floats manufactured by themselves; thus they would drift along as plankton with the currents, and are world wide in distribution. This makes them highly useful as guide fossils, even in the correlation of deposits at great distances from each other. Most graptolites are found in black shales, where they appear as flattened, calligraphic markings. Occasionally, however, they may also be preserved in limestone, from which they can be etched and studied in detail by means of serial sectioning and other techniques. As a result, we now have detailed information on the anatomy of these animals.

The relationships of graptolites have been much discussed; according to one school of thought, they may be related to the present-day pterobranchs (a group of small colonial animals belonging to the same broad division as the vertebrates). However this may be, the graptolites may probably still be regarded as a phylum of their own.

While graptolite shales were formed in the geosynclines, limestones with trilobites, orthoceratites, and other fauna were deposited in the shelf seas. Among Ordovician trilobites may be noted the asaphid family with forms like *Asaphus* (a trilobite known to Linné) and the giant *Megistaspis*; these forms have a large pygidium and 8-10 body segments. *Illaenus* was a very smooth, highly arched form; *Ampyx* a small, almost disc-like trilobite with three enormously long spines extending forward and backward from the head shield; *Trinucleus* another small form with enormously expanded cheek region and spines in the head shield. *Staurocephalus* is remarkable for the globular development of its glabella.

Another group that became dominant in the Ordovician was the nautiloid cephalopods, which survive today in the pearly nautilus. Their Ordovician representatives mostly had a straight shell, and these orthoceratites may sometimes cover limestone surfaces in great profusion. Partly coiled shells are also known.

Several other groups yield important fossils, for instance the echinoderms, represented by such forms as the Cystoidea – globular or elongate, sedentary animals; and the brachiopods, which were now extremely varied and numerous. In the Ordovician we also find the first evidence of vertebrates, unfortunately only as small bone fragments, perhaps carried by the rivers out to sea from the original habitat in lakes and streams. The bone structure, however, closely resembles that found in primitive vertebrates later on.

By the end of the Ordovician, all of the living animal phyla that have a fossil record at all had appeared in the record. They are the Protozoa, or one-celled animals; the Porifera, or sponges; the Coelenterata or hydroids, medusae, and corals; the *Annelida* or segmented worms; the Bryozoa or Polyzoa, 'moss animals'; the Brachiopoda, or lamp shells; the Mollusca; the Arthropoda; the

Chaetognatha, or arrow worms; the Echinodermata; and the Chordata, with the vertebrates. The basic diversity of the modern animal world is thus of very long standing.

At the same time, remains of less successful phyla were still straggling on. The graptolites are of course a special case, for they reached their acme in the Ordovician. But there are other, small, problematical groups of unknown affinities, such as *Hyolithes* already described, and *Receptaculites* – a semiglobular, somewhat flattened organism sometimes reaching the giant dimensions of two metres in diameter. It shows some resemblance to sponges but is clearly not a sponge. All of these phyla, the successful and unsuccessful alike, probably arose by radiating evolution well before the Cambrian.

In North America, a phase of orogeny called the Taconian resulted in the uplift of a highland east of Pennsylvania and New York; remnants of the Ordovician mountains are now seen in the Taconic Mountains of eastern New York.

The Ordovician period probably lasted some 70-80 million years.

## The Silurian

The Silurian period brought the long years of calm and quiescence of the Cambrian and Ordovician to an end in Europe. The late Ordovician to early Silurian initial orogenic phase was followed in the late Silurian by the main Caledonid revolution, which continued into Devonian times and resulted in the folding and uplifting of the Scandinavian and Scottish highlands. In North America there was no renewal of mountain-building in the Silurian, but the Taconic highlands that were produced in the Ordovician continued to feed sediments into the eastern geosyncline.

In Silurian deposits we find the earliest well preserved fossils of vertebrates. The great majority belong to the Cyclostomi or jawless 'fish', which now survive only in the lampreys and hagfishes. These modern cyclostomes have a cartilaginous skeleton, but their Silurian and Devonian relatives had a heavy bone armour and are

17 Trilobites are still important in the Ordovician, and highly varied in appearance: a, *Asaphus*; b, *Cryptolithus* (a trinucleid); c, *Staurocephalus*; d, *Ampyx*; e, *Illaenus*. Reduction and even loss of eyes is fairly common (b, d, e). Nautiloid cephalopods are also very common, especially the straight-shelled orthoceratites, but also partly coiled forms like (f) *Lituites*. (Not to scale.)

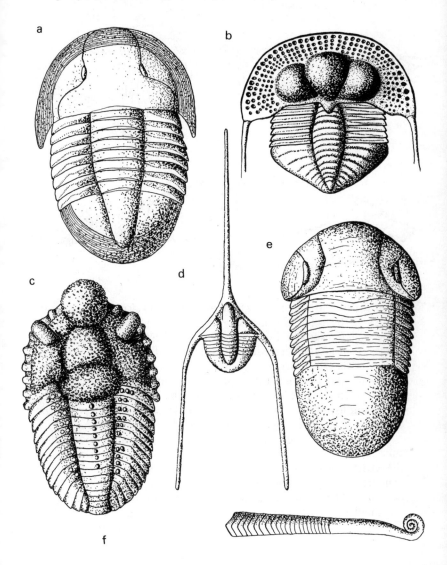

18 Ostracoderms, or jawless fish of the palaeozoic, are the first vertebrates to appear in the fossil record; the Silurian was their heyday. Shown here are representatives of their two main divisions: (a) the cephalaspidomorph *Hemicyclaspis*, and (b) the pteraspidomorph *Anglaspis*. Both have a large head shield and scale-clad body, but the details are quite different; note the inverse structure of the tail fin in the two forms, and the remarkable lateral field along the cheek of *Hemicyclaspis*.

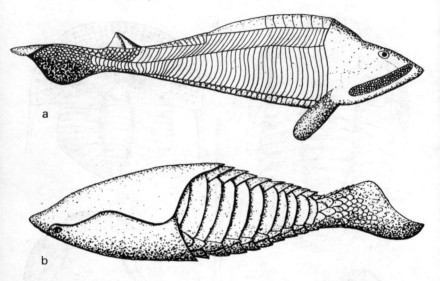

often called ostracoderms. There are two main groups of ostracoderms in the later Silurian. The earliest come from the Ludlow bone bed in western England, while the best preserved material comes from late Silurian deposits (often reddish sandstones transitional to the Old Red) in England, Scotland, Norway, Spitzbergen, Esthonia, and Podolia. Intensive studies have been made of these jawless forms, especially by means of the serial sectioning technique employed by E. A. Stensiö and his school, and as a result the anatomy of these early vertebrates is known in remarkable detail.

The best-known group is termed the Cephalaspidomorphi; it survives in the living lamprey. The Silurian and Devonian forms, the earliest of which appear in the Ludlow, were mostly less than $\frac{1}{2}$ metre in length. The head was completely encased in a bony carapace. The eyes were closely spaced on top of the head, while the small jawless mouth opening lay ventrally near the front end. The body, which was flat-bellied and triangular in cross section,

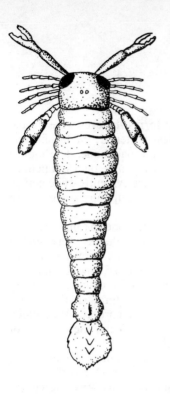

19 Largest arthropods of all times were the eurypterids. Some species of the genus *Pterygotus* shown here were more than six feet long, others only a few inches. Structure of eurypterids suggests that they swam on their backs like many aquatic insects, using hind legs in oar fashion, also flapping the movable plates on the upturned belly to aid rapid propulsion. Large eyes and powerful claws indicate predaceous habits of this form. Other eurypterids have a spike-shaped tail.

was clad in large scales; the big tail fin was supported along its upper margin by the backbone. One or two fins were present on the back, and two flaps somewhat resembling pectoral fins were attached to the hind edge of the head shield, which might also carry a pair of large recurved spines. The large mouth cavity of these creatures carried ten pairs of internal gills, and these were probably used to strain food particles from the bottom mud; these ostracoderms were evidently bottom dwellers. Peculiar lateral fields are seen in most forms along the cheeks, and a similar central field behind the eyes; these are probably tactile sense organs of some kind.

A second main group of ostracoderms, the Pteraspidomorphi, appeared as early as the basal Ordovician (fragments from Esthonia) and is also found in later Ordovician strata, but well-preserved material is only known from the Silurian and Devonian. Like the cephalaspids they had a head shield, which may encase part of the body as well. But in these forms the eyes were on the sides of the head, and the body was less flattened; the tail fin was supported along its nether margin by the backbone. The mouth forms a transverse slit in front with some movable plates, functioning in the manner of jaws. These forms were evidently quite active, perhaps part-predaceous swimmers. The group survives today in the hagfish.

Though the Silurian and Devonian ostracoderms may be considered primitive as vertebrates, they nevertheless were highly specialised in many respects. Clearly the forms known to us are not directly ancestral to the jawed vertebrates, and it must be assumed that the jawed and jawless (fish) had a common ancestry well back in Cambrian or even Precambrian times. If these early vertebrates lived in freshwater, as the fossil record and the sudden marine invasion in the Silurian seem to indicate, the chances of finding any fossil evidence are very slim indeed.

Trilobites continued to flourish in the Silurian with many new, remarkable types (including some very spiny ones), but the largest and most powerful animals of the Silurian seas were the eurypterids. This group is known from the Ordovician to the Permian but reached its apogee in the Silurian and Devonian. They were elongated arthropods, up to six feet long, with a relatively small head shield, many body segments, and a long tail spike. Many of them were armed with powerful, pincer-like claws, and were evidently predaceous; they had two well developed compound eyes, looking almost straight forward in the most carnivorous forms, and in addition two small simple eyes on top of the head. The eurypterids seem to have shunned the open sea and may have preferred coastal lagoons and estuarine habitats. No doubt they posed a serious danger to other animals in their life zone, including the early vertebrates.

Among Silurian arthropods are also found scorpions, which may be the first air-breathing animals in the geological record, and millipedes or 'thousand-legged worms'. That a suitable habitat for land animals was at long last becoming available is shown by the presence of true land plants. These Silurian land plants are (like their Cambrian predecessors) related to the modern club mosses (*Lycopodium*); they form branching stems with sporangia at the base of the leaves. The evolution of a land vegetation was now well under way, though true forests were yet to come.

In the Silurian geosynclines, graptolite shales were still being deposited, while the continental seas produced limestones and sandstones. Coral reefs were now becoming common; these corals belong to other, more primitive groups than those of the present day. Cephalopods, both straight-shelled and coiled, were still numerous, while amongst echinoderms the crinoids or stone-lilies were crowding out the cystoids.

Some small isolated phyla were still holding out in the Silurian – the *Receptaculites* group and the *Hyolithes*, and also the Conulariida, a phylum of uncertain affinities that ranges in time from the early Cambrian to the Trias. *Conularia* had a pyramid-shaped, chitinophosphatic test, the basal aperture of which could be closed by four triangular flaps bending in from the side walls. Different authors have suggested relationships to practically all the known animal phyla except the protozoans, sponges, and echinoderms!

The Silurian appears to have been the shortest of all the geological periods (except the Quaternary), and lasted only about 25-30 million years.

## The Devonian

Heralded by the ostracoderms of earlier periods, the vertebrates appear in a profusion of fish-like forms in the Devonian. This period has in fact been called the Age of Fishes, and is estimated to have lasted some sixty million years. From the late Devonian we already have evidence of four-footed vertebrates or Tetrapoda.

The Devonian in Europe is the time of the great Old Red Continent (sometimes called Palaeoeurope) that extended from England to the sea covering the Ural geosyncline. The red beds were deposited along the edges of this continent, in a climate thought to have been dry and warm; inland there may have been great deserts. Along the south coast of the continent, the deepening Variscan geosyncline extended from west to east, being bordered to the south by a large island extending from central France eastward. Southern Europe was almost completely inundated. The end of the Devonian is marked by the beginning of the great Variscan orogeny.

In North America the Devonian was a time of great unrest, concentrated in the northern part of the eastern geosyncline, and beginning in the Middle Devonian. This orogeny, termed the Acadian, resulted in the formation of a mountain range in the east, and so the relief of this continent was again rejuvenated after the wearing down of the Taconian range that had been born almost a hundred million years earlier.

The animal life of the Devonian seas has come down to us in great richness and variety, but the most interesting creatures are the vertebrates. Besides the ostracoderms – which survived almost throughout the period – we now find many kinds of (true) fishes, of which only a few precursors have been recorded in the Silurian rocks. Particularly good material has been brought back by expeditions to Greenland and Spitzbergen, but the Old Red Sandstone in England and elsewhere has also yielded many fine specimens. The serial sectioning technique of the Stockholm school has been of special importance in these studies.

The largest predators of the time belonged to the arthrodires or joint-necked fish, of which the giant *Dinichthys* was almost ten metres long. The bony armour of the head was connected with that of the fore part of the body by means of a kind of hinge-joint. But the arthrodires, which appeared in the Silurian and reached their apogee in the Devonian, also include many smaller forms, of great variety. One of these groups still survives in the form of the peculiar present-day chimaeras or ratfish. Interestingly, there may also be some special relationship between the arthrodires and the lungfish or Dipnoi, which also made their first appearance in the Devonian. The lungfish of today still closely resemble their Devonian ancestors.

More orthodoxically fish-like than the arthrodires were the acanthodians or spiny-finned fish, which range in time from the late Silurian to the Permian, but were most varied in the early Devonian. True sharks appear in the Middle Devonian, together with the first ray-finned fish; between them these two groups still form the great majority of the fish fauna. But the Devonian forms, of course, were different in many respects from their modern descendants.

Still another group of fishes is even more important to us, for it includes the ancestors of all the land-living vertebrates (or tetrapods), including ourselves. These were the Crossopterygii or lobe-finned fishes. There are two main divisions, and their subsequent histories show the most remarkable contrast between conservatism and progress in evolution.

The conservative group, the Coelacanthini or fringe-finned fish, was destined to remain water-living. It gradually petered out during the course of the Palaeozoic and Mesozoic, and vanishes from the geological record at the end of the Cretaceous; the coelacanths were long thought to have become extinct. But in this century, living coelacanths very like their earliest fossil precursors, have been taken off the coast of South Africa, to provide a famous instance of a 'living fossil'.

The second crossopterygian group, the Rhipidistia, does not survive as such, but instead gave rise to all the land-living vertebrates, so that its history is one of unmatched evolutionary success. The rhipidistian fishes resemble the earliest amphibians in many respects: the bones of the skull, the rather peculiar structure of the teeth, and the arrangement of the bones in the paired fins, which is similar to that in the tetrapod limbs. Probably these fishes already used their fins for movement on dry land. A. S. Romer has suggested that they moved from pool to pool in an arid climate that might lead to seasonal drying of the smaller pools. Thus, the lungfish might withstand a drought by burying themselves in the

20 Devonian fishes. (a) *Pteraspis*, an ostracoderm; (b) *Climatius*, a spiny acanthodian; (c) *Eusthenopteron*, a member of the Crossopterygii, or lobe-finned fishes, close to the ancestry of land vertebrates. Lobes at fin bases contain jointed skeleton which gave rise to the limb bones of the tetrapods.

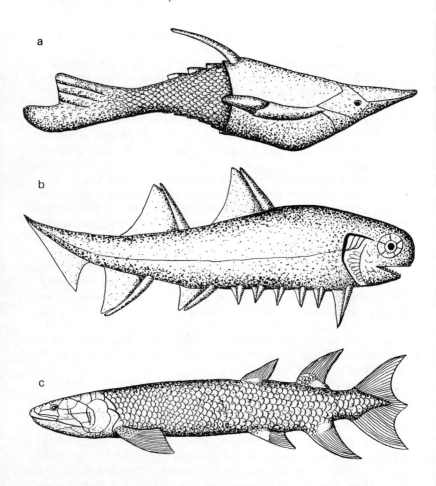

21 Devonian fishes and amphibian. (a) *Coccosteus*, an arthrodire or joint-necked fish; (b) *Rhamphodopsis*, possibly ancestral to present-day chimaeras; (c) *Bothriolepis*, a member of the Antiarchi, peculiar forms with 'arms' encased in a jointed exoskeleton; (d) *Ichthyostega*, the earliest-known amphibian, with fish-like tail and weak, primitive limbs.

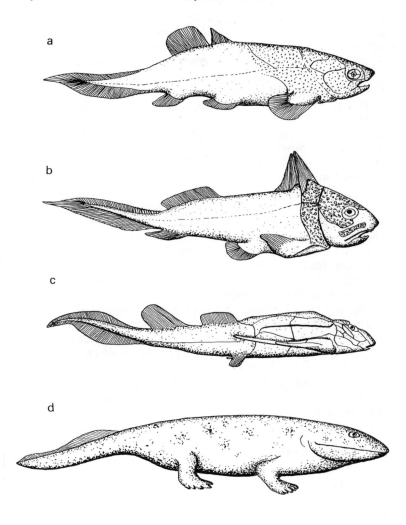

moist bottom mud, while the rhipidistians responded in a more active way, by moving off in search of other waters. The presence of internal nares indicates that these fishes possessed lungs.

The first amphibians appear towards the end of the Devonian. Both footprints and actual skeletons have been found, the latter in eastern Greenland. These early amphibians still resemble the rhipidistians in many features, down to the fish-like tail, but the limbs are unmistakably of the tetrapod type. It would seem that the transition from fish to amphibian took place in the Middle Devonian, probably within several different groups of rhipidistians independently of each other.

Many kinds of land plants are known from the Middle and Upper Devonian. The most primitive were simple forms looking somewhat like a kind of seaweed; but ferns, seed ferns, horsetails, and scale trees also appear as forerunners of the majestic forest flora that was to appear in the next period.

In the seas, brachiopods were especially important among invertebrates, and make useful guide fossils. Primitive echinoderm types were still much in evidence, but starfish and sea-urchins were on the increase. Devonian coral reefs are widespread. Trilobites and graptolites were still in existence but less prominent than in earlier times, and among cephalopods the nautiloid group was now giving way to the first ammonoids – a group to be discussed in more detail later on. The big eurypterids continued to hold their own in fresh and brackish waters, while the earliest insects (primitive forms related to the living Collembola or Springtails) appeared in the Middle Devonian.

## The Carboniferous

If the Devonian was the Age of Fishes, the 60-65 million years of the Carboniferous period may well be styled the Age of Amphibians. In this period the tetrapods multiplied greatly, and soon the lowlands were populated by great swarms of land animals. It is unlikely, however, that they would have ventured very far from streams and lakes, for amphibians remain dependent on water for their reproduction and early stages of life.

The Carboniferous in Europe is the time of Variscan mountain-building. The Variscan geosyncline gradually filled up and arose to form the new mountain ranges, while coal was formed on the coastal plains of the ancient Old Red Continent and along the borders of the geosyncline, where great swamp areas were created and an impressive forest vegetation developed. In North America the Carboniferous is divided into two periods, the Mississippian and Pennsylvanian. The Variscan orogeny of Europe had its counterpart in North America in the uplifting of the old Colorado Mountains.

The Coal Measures represent the fossil remains of the Carboniferous forests, the great majority being of late Carboniferous, or Pennsylvanian, date. The swamp forests consisted of several types of trees, of which the great scale trees (*Lepidodendron*) were up to 100 feet high. The 'scales' are in fact leaf scars, present on the surface of the trunk and branches alike. The related *Sigillaria* was not quite as tall, and had larger leaves. The next lower level of the forest is formed by giant horsetails or scouring rushes (*Calamites*, etc.), which must have formed dense thickets. Finally, the undergrowth consisted of large seed ferns and true ferns.

The drier ground above the swamps supported a different type of forest, characterised by leafed conifers (*Cordaites*, *Walchia*) and ginkgo-like trees.

There was a rhythmic pattern of transgressions and regressions so that the lowlying coal swamps were often overrun by the sea and covered by a blanket of marine sediment. Under this cover, the organic matter was gradually turned into lignite and ultimately into coal. This conspicuous rhythm is probably connected with the formation of continental ice fields in the southern continents. At this time the South Pole was within Gondwanaland, producing a protracted Ice Age. As these ice fields waxed and waned – just as those of the Quaternary Ice Age were to do much later – water was taken up from the oceans, or returned to them. The coal swamps,

22 The primitive amphibian *Eryops* from the early Permian Red Beds of Texas. The sprawling stance is like that of a modern salamander, but *Eryops* was from five to seven feet long and weighed several hundred pounds, so that its movements on land can only have been very slow and cumbrous; presumably these animals spent much of their life in water. Rhachitomous amphibians of this type were common in the later Carboniferous.

of course, were far removed from the glaciated areas, in a persistently tropical climatic zone.

Their animal life was rich and striking. There were several kinds of insects, related to cockroaches, may-flies, and dragon-flies; many of them were very large. The greatest insect known was a Carboniferous dragon-fly with a wing span of up to 70 centimetres. If its nymphs resembled those of modern dragon-flies, they must have been remarkable predators of the swamp lakes. An interesting insect group, the extinct Palaeodictyoptera, retained an extra pair of vestigial wings in front of the two ordinary pairs. Other land arthropods include spiders, scorpions, and centipedes.

The amphibians of the Lower Carboniferous (Mississippian) are sometimes called stegocephalians, a name referring to the solid, unfenestrated roof of the big flat head, still resembling that of the rhipidistians in many features. The early stegocephalians possessed comparatively weak, sprawling limbs, which would hardly have been able to lift the heavy body for any length of time; so they probably used their limbs to push and pull themselves forward, in moving from one pool to another. Their mode of life probably was basically the same as that of their rhipidistian ancestors, for only one form of the latter survived into the early Carboniferous, and that soon became extinct too.

Most later Carboniferous land-living amphibians belong to the Rhachitomi, which were fairly strong-limbed, powerful creatures with enormous, tooth-studded jaws; they were carnivorous, which was the rule in early amphibians. Other amphibians tended to return to the water for good. The giants among these are called embolomeres; they had weak limbs but powerful swimming tails, and probably resembled crocodiles in habits. Various other water-living amphibians inhabited the coal swamps; some of them were snake- or eel-like in appearance, with reduced limbs or no limbs at all. A group of very odd-looking bottom dwellers are the diplocaulids with enormous, flat, wedge-shaped heads.

Other amphibians show closer resemblance to reptiles, and in the Upper Carboniferous the first true reptiles are already on the scene, although we do not yet have direct evidence that their eggs were of the reptilian, shelled type. The earliest known reptile egg dates from the early Permian. It is probable that these more agile animals inhabited drier ground away from the coal swamps.

In the seas of the Carboniferous, crinoids or stone-lilies were among the dominant invertebrates, together with brachiopods belonging to the spiriferid and productid groups. Ammonoids were becoming abundant, while the trilobites were now declining, so that there were only two genera left in the late Carboniferous. Sharks and ray-finned bony fishes were already completely dominant in the fish fauna, while the more primitive fish groups were becoming extinct.

The foraminifers of the Carboniferous include a remarkable group, the giant fusulinids, which were especially common in the late Carboniferous seas; some rocks of this age are made up almost exclusively of these one-celled, grain-shaped animals.

The Carboniferous saw the last of the graptolites, eurypterids, and *Receptaculites*.

23 *Pareiasaurus*, a giant cotylosaurian reptile from the Permian of the Karroo, Africa; length ten feet. As restored, the skeleton shows an early stage in the development of true walking, with limbs flexed at ninety degrees – a mechanically inefficient design necessitating excessively powerful muscles and limb bones. The backbone should be more arched than in this mount.

## The Permian

The Permian period, which lasted some 50 million years, brought the Palaeozoic Era to a close. It was a period with continuing orogenic disturbances, and with cold conditions carrying over from the late Carboniferous glaciation of the southern hemisphere. In this period fall the last phases of the great Variscan orogeny in Europe; the folding and uplifting of the Ural Mountains; the elevation of the Appalachians in eastern North America; and various other local orogenies in Asia, Australia, and South America. The Permian indeed was a time of exceptionally high relief of the face of the earth, just like the present, and with regressive seas. It also seems that the initial cracking up of the ancient Gondwana Continent was taking place in the Permian, perhaps along rift lines more or less like the modern Rift Valley of East Africa and Palestine.

Famous continental beds, with rich fossil deposits, are known from the Permo-Triassic of the Karroo of South Africa, the Permian deposits of Russia, and the Red Beds in Texas and New Mexico. The southern continents carried a highly characteristic, cool-climate vegetation, the *Glossopteris* flora. These plants, which resembled ferns but were not closely related to them, formed veritable forests, and some have given rise to coal fields. Meanwhile, the northern continents were becoming too cool and dry for the Carboniferous-type swamp forest to survive except in local basins; the dominant forest trees were now primitive conifers and ginkgos.

Many amphibians very like their Carboniferous ancestors survived in the Permian, for instance the Rhachitomi on land and the diplocaulids in the lakes. But dominance on land had already passed to the reptiles; indeed it may be said that the Age of Reptiles begins with the Permian period.

The ancestral reptile group is called the Cotylosauria, or stem reptiles; it dates back to the Carboniferous. The earliest cotylosaurs were small and medium-sized carnivorous forms, but in the Permian herbivorous cotylosaurs made their appearance; these

were the first plant-eating Tetrapoda. They are easily identified by their peculiar, leaf-shaped or peg-shaped teeth, quite dissimilar from the pointed fangs of the flesh-eating reptiles. The largest of these forms were the big, lumbering pareiasaurs of South Africa and Europe; they measured up to 10 feet in length and in the flesh may perhaps have looked a little like hornless rhinoceroses.

A closely related group, the Pelycosauria, also contains both flesh-eating and plant-eating types. Some of these animals (both herbivorous and carnivorous) had greatly elongated spines on their backs, probably supporting a big skin fold not unlike the dorsal fin of the living sailfish. The function of this bizarre-looking structure has been much debated; the most reasonable guess seems to be that it was a heat-radiating device to help control the body temperature. Some early Permian pelycosaurs probably gave rise to the Therapsida, or mammal-like reptiles, which became common in the later Permian.

The therapsids, again, evolved into both carnivorous and herbivorous forms. Their dentition was becoming differentiated into cheek teeth for chewing, big eye-teeth or canines for attack and defense, and small front teeth or incisors for nipping. In one

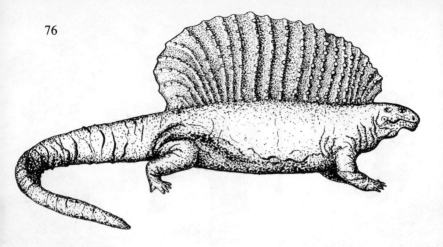

group called dicynodonts, all the teeth except for the upper canine tusks were lost, and even these could be missing; the jaws probably were covered by a horny beak. The dicynodonts were herbivorous, whereas the contemporary gorgonopsians were wicked-looking carnivores with great sabre-like canine teeth. The largest of all therapsids and indeed of all Permian land animals were the dinocephalians, which again include both meat-eating and plant-eating forms.

Even the earliest reptiles exhibit a trend back to the water, as evinced by *Mesosaurus* from the earliest Permian; this small reptile looked somewhat like a minute, bristle-toothed crocodilian.

Permian insect life shows much greater diversity than that of the Carboniferous. Dragonflies and cockroaches of primitive types continued to flourish, but in addition there are now grasshoppers, lacewings, book-lice, plecopterans (stone-flies), mecopterans (scorpion flies), true beetles and some extinct beetle-like orders, and finally primitive true flies or dipterans.

Sharks and ray-finned fishes continued to be successful in the Permian seas, but a number of ancient groups made their last appearance. For example the trilobites, the *Hyolithes* group, and the blastoid echinoderms, all of which have a history going back to Cambrian or Ordovician times. Other forms were becoming rare, and would face extinction shortly after the Permian. For instance, the corals were declining seriously; probably the cool climate of the period was now helping to weed out the ancient coral types (the Tabulata and Rugosa) that had been so successful in the

24 The long-spined, plant-eating reptile *Edaphosaurus*, a member of the Pelycosauria (primitive mammal-like reptiles); length eleven feet. The spines, which in this form carried small crossbars, supported a skin fold that produced an increase of the body surface relative to its volume. This character is lacking in smaller, related forms. The relationship between surface and volume is important in the heat economy of an animal, which suggests an adaptive significance of the 'sail'.

earlier Palaeozoic. The modern hexacorals evolved at about this time, but were still relatively scarce. Again, the Permian was a crucial time for the brachiopods. Several groups that had been dominant in the Palaeozoic seas now died out at the close of the Permian, and only a few types of brachiopods survived into the Mesozoic; indeed they are all still with us, except for the spiriferids.

The big fusulinid foraminifers also became extinct at the end of the Permian, despite the fact that they flourished greatly throughout the period. Among cephalopods, ammonoids were now very numerous, and the suture lines formed by the chamber walls were becoming increasingly complicated; the 'goniatitic' simple suture of early ammonoids was being transformed into a more lobate, 'ceratitic' type.

## The Mesozoic

We now enter the Triassic period; the stage is being set for the dinosaurs. Many new and remarkable forms were to appear in the 50 million years of Triassic time, among them the first mammals and the first dinosaurs. This took place in a world that was gradually returning to its normal state from the unrest and upheavals of the Permian. Compared with the late Palaeozoic, the Trias was a serene period.

But at the same time the seeds of future unrest were sown, for in the Trias the great Tethyan geosyncline system was developed. Out of this system, the Alps, Himalaya, and other young mountain ranges were to evolve two hundred million years later.

The Trias was named in Germany where the Triassic system is formed by three successive series of rocks, corresponding to as many epochs. The lowermost and thus oldest is the Bunter Sandstone with vividly coloured sediments laid down in shallow depressions. There follows a limestone series with marine shells, the Muschelkalk; and finally a series of continental deposits, the Keuper, with beds of rock salt and gypsum, indicating a very dry climate. Later, a fourth series, the Rhaetic, was included in the Trias.

The Triassic marine faunas show profound changes relative to those of the preceding Permian; this is one of the great turning points in the fossil record. The fusulinid foraminifers were gone. The Palaeozoic corals were also becoming extinct, and were superseded by the new hexacorals, which became very common in the Tethys. Of the typical Palaeozoic brachiopods, only the spiriferids remained, and it would seem that molluscs – gastropods and bivalves – were taking over many former brachiopod niches. Among the ammonoids, the simple-sutured goniatites were replaced by complex-sutured forms. The related, squid-like belemnites appear in great numbers in the Triassic seas; precursors of this group are however known from the late Palaeozoic. The trilobites were gone, and so were the other long-lived Palaeozoic groups except for the Conulariida, which persisted into the Triassic, then became extinct.

There is more continuity in the land faunas of the Trias, which are known from many parts of the world – for instance southwestern United States, South Africa, and India; an interesting range of small forms has been collected in European fissure fillings of Rhaetic age. The lowliest of the tetrapods, the stegocephalian amphibians, continued to flourish throughout the Triassic, and make their last appearance in the Rhaetic. Among reptiles, both

25 Large predaceous pelycosaurs like *Dimetrodon* also tended to develop a sail-like dorsal skin fold, evidently as a reaction to the same need for a constant surface/volume relationship. The existence of such structures in the Pelycosauria may suggest a foreshadowing of the higher rates of activity and metabolism that were to characterise the mammals.

cotylosaurs and therapsids carried over into the Trias, and the latter developed some very advanced forms. An even more progressive group of mammal-like reptiles, the Ictidosauria, succeeded the therapsids in the later Trias, and by this time true mammals were also in existence.

Many new reptile types appeared in the Trias. Here, for instance, we meet the first turtles, basically like their living descendants but still with teeth in their jaws; all modern turtles are toothless. Even more important was the appearance of the first archosaurs, at first in the form of thecodonts, later on as true dinosaurs. There is also a host of swimming forms: ichthyosaurs or fish-lizards, nothosaurs (related to the plesiosaurs that will be described later on), and the aberrant, flippered placodonts, highly specialised swimming reptiles with big flat teeth presumably adapted for the crushing of mollusc shells. Many of these animals will be described in more detail below.

The Triassic land flora is dominated by coniferous trees, best exemplified by the majestic logs of *Araucarioxylon* in the Petrified Forest of Arizona; there are also ginkgos, cycads, and bennettitales. Locally in swampy areas, the ancient *Lepidodendron* was still to be found, and big ferns and horsetails remained plentiful.

The remaining two periods of the Mesozoic, the Jurassic and Cretaceous, will be the main topic of the remainder of this book. However, by way of introduction, a résumé of their chronology will be given here.

The Jurassic, which probably lasted about 40 million years, was named after the Jura Mountains. Like the Trias, this system of rocks may be divided into three parts. The lowermost is the Lias (Black Jura of German usage), upon which follow the Middle Jurassic (Dogger or Brown Jura), and the Upper Jurassic (Malm or White Jura). In the Jurassic of Europe most fossils are of marine animals, for a large part of the continent was then submerged. The marine fauna of the Liassic epoch is well represented in Yorkshire, on both sides of the English Channel, and in southern Germany, where Holzmaden in Württemberg is a famous site with

## 2 The Mesozoic, showing the Jurassic and Cretaceous periods, and important formations and localities for fossil vertebrates

| Period | Epochs with stages | Formations yielding fossil vertebrates in Europe | elsewhere |
|---|---|---|---|
| Cretaceous | Upper Cretaceous: Senonian Turonian Cenomanian | The Chalk (*England, France, Belgium, Westphalia,* etc.) | Lance, Edmonton, Belly River (continental) Pierre, Niobrara, Benton (marine) Djadochta |
| | Lower Cretaceous: Albian Aptian Neocomian | Greensand, Gault (*S. England*), Neocomian (*Istria*), Wealden Beds | *Cloverly* (Wyoming), *Mongolia, Shantung, Australia* |
| Jurassic | Upper Jurassic (Malm): Purbeckian Portlandian Kimmeridgian Oxfordian Callovian | *Solnhofen,* Purbeck, Oxford Clay | Morrison, *Tendaguru, Sundance* |
| | Middle Jurassic (Dogger): Bathonian Bajocian | Stonesfield Slate | Navajo Sandstone (*Arizona*) |
| | Lower Jurassic (Lias): Toarcian Pliensbachian Sinemurian Hettangian | Holzmaden, Yorkshire | Arizona, Australia |
| Trias | Keuper Muschelkalk Buntsandstein | | |

exquisitely preserved fossils. The Upper Jurassic is represented in Germany by another celebrated site, Solnhofen, where the extremely fine-grained sediments have preserved a wonderful array of organisms. In England, the Upper Jurassic Oxford Clay contains rich reptile and fish faunas, while Stonesfield and Purbeck give some glimpses of the contemporary land animals. In North America the late Jurassic Morrison beds have yielded a remarkable dinosaur fauna, and somewhat similar deposits of the same age are known from Tendaguru in Tanganyika.

The Cretaceous period is the time when the famous Chalk deposits were formed. In contrast with the Trias and Jurassic, the Cretaceous is primarily divided into two epochs only, despite the fact that it was a very long period – probably upward of 80 million years. From the earlier Cretaceous, fossil land animals are primarily known from the Wealden beds in England and Belgium, and other freshwater deposits especially in east Asia, while marine faunas are widespread, for this was a time of high sea levels. Still, rich land faunas are also known from the Upper Cretaceous, especially in North America and Mongolia. There is a classic Upper Cretaceous sequence in Canada and the United States, containing the following succession of stages: Belly River, Edmonton, and Lance. In Mongolia, the Upper Cretaceous Djadochta beds have proved to be richly fossiliferous, and the deposits in this area are now being investigated by Mongolian, Russian and Polish scientists.

# 3 The dinosaurs

## The origin of dinosaurs

The earliest archosaurs are grouped in the Order Thecodontia (the name alludes to the fact that the teeth were implanted in alveoli in the jawbones, a characteristic of all archosaurs). Thecodonts existed throughout the Trias. Most were small reptiles up to three feet in length, but they differed from modern lizards in that the hind legs usually were much longer than the front; many were probably bipedal. This was an important innovation among the land animals of that time and may have been a key to the success of the archosaurs for so many million years. The bipedal tendency is indeed a leading motif among the ruling reptiles.

The earliest reptiles possessed very primitive, sprawling limbs, which were not efficient mechanically. The upper arm and thigh protruded horizontally from the clumsy body, and to raise that body from the ground required a great effort. The arrangement functions very well in small animals like salamanders and small lizards, but with increasing weight it becomes very cumbersome. The primitive reptiles, for instance the pareiasaurs, were often big and heavy, and so their movements must have been toilsome, slow and shuffling.

In more advanced reptiles the limbs tended to be swung in under the body, functioning in the manner of pillars rather than levers; the belly was raised above the ground, and these reptiles invented the art of walking. The rule of the land gradually passed to the therapsids and other advanced reptiles with this type of gait. Most were four-legged, but among thecodonts the front legs tended to become relatively short, and a completely new mode of standing and walking appeared. These reptiles, somewhat like the kangaroo, have long legs, short arms and a long, powerful tail, which balanced the body so that the centre of gravity came to lie at the hips. Thus the reptile would stand up on its hind legs, but it apparently did not move by hopping in kangaroo fashion, but by running and walking. When running the body was probably poised horizontally, balanced by the raised tail, which probably swung

26 Locomotion in reptiles. Primitive forms like *Pareiasaurus* (a) retained essentially the salamander-type sprawling limbs, acting like levers to raise the body from the ground. In advanced forms such as the mammal-like *Cynognathus* (b) limbs were swung in under the body to support it, pillar-like. Finally, in archosaurs like *Hypsilophodon* (c) stance becomes bipedal, with body semi-erect and balanced by the tail, leaving arms free for other functions.

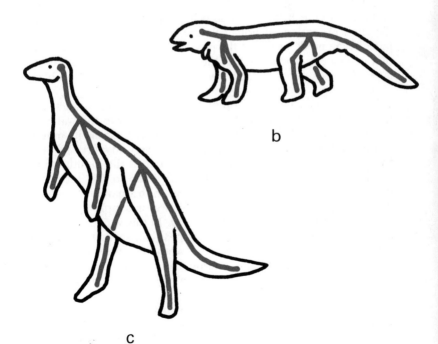

lightly from side to side with each stride. The step from the functionally four-legged type, as shown in the figure on the previous page, is not very long.

When resting the animal supported itself on its hind legs and tail, while the body was reared up: this gave a stable tripod position, as well as a higher vantage point from which to survey the surroundings. From thecodonts of this general type, all the later archosaurs seem to have evolved; perhaps from forms not unlike the *Ticinosuchus* shown here.

A quite different group of thecodonts is formed by the suborder Phytosauria, which switched to life in the water and came to resemble the crocodiles very closely. (The name means plant-reptile and is due to the mistaken initial belief that they were herbivorous; the rules of zoological nomenclature do not permit a change of name.) The mode of life of the phytosaurs was probably very like that of modern crocodiles, but they were not true ancestors to the latter. It turns out that they have solved the 'breathing problem' in a completely different way.

In an ordinary land-living reptile the nostrils open straight into the mouth cavity. If a reptile like this ventures into the water, its breathing will be blocked by water as soon as it opens its mouth. The crocodiles have solved this problem in a very advanced manner, by evolving a secondary palate or false roof to the mouth, which forms a closed air passage all the way back to the throat; in principle this is the same arrangement as in the mammals. In most water-living reptiles, however, the solution is simply to shift the nostrils back toward the top of the head. In this way they come as close to the throat and windpipe as possible, so that the animal can open its mouth when submerged without getting water down the wrong way. This was the solution in the phytosaurs, and we are going to meet the same arrangement in many other reptiles, certain dinosaurs for instance.

The two dinosaur orders, the Saurischia (reptile-hip) and Ornithischia (bird-hip) differ in the build of the pelvic girdle. In the saurischians the triradiate build of the thecodont hip girdle was

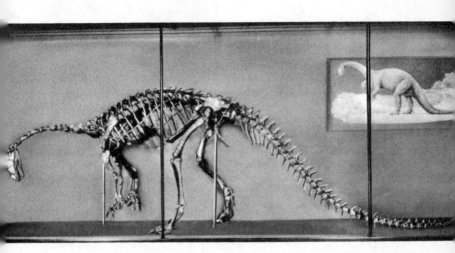

27 The herbivorous *Plateosaurus* (length twenty feet) of the late Trias in south Germany is typical of the Palaeopoda, a group that may be ancestral to the gigantic sauropod dinosaurs of the Jurassic and Cretaceous. The resemblance is seen in the small head, long neck, and long tail; unlike the sauropods, however, *Plateosaurus* was able to rear up on its hind legs.

retained – as it also is in the crocodiles. The hipbone or ilium is horizontal, while the pubic bone extends obliquely downward and forward, and the ischium downward and backward. This arrangement gives excellent leverage for the powerful muscles that move the legs forward and backward.

In the ornithischians the pubis points downward and backward and is parallel to the ischium (as in the birds). At the front end of the pubic bone a new bony process is formed, pointing forward and downward. This type of girdle, then, is tetraradiate.

Members of the order Saurischia are quite common in the upper Trias (Keuper), while ornithischians are scarce in the Keuper and appear in greater numbers only in the Jurassic.

### How much did a dinosaur weigh?

When we look at one of the mounted skeletons of an extinct giant reptile, we may perhaps wonder how much the creature weighed in life. This can be calculated with some precision, and a systematic

series of estimates has recently been made by E. H. Colbert. The method is as follows.

The estimates are based on miniature restorations of the dinosaurs, as they are assumed to have been in the flesh. A restoration starts with the mounted skeleton. Study of the bones will show how the musculature was distributed, so that the skeleton may again be clad with muscles and an approximate body outline may be produced. The occasionally preserved skin impressions may give an idea of the external aspect of the body. Studies of this kind enable us to build up a picture or sculpture giving a general idea of what the dinosaur looked like. There are of course some details we can never know, as regards the colour, skin folds, etc.

Colbert was now able to measure the volume of a miniature sculpture of the dinosaur. By ascertaining the scale of the model it was then possible to estimate the volume of the original dinosaur.

28 Bones of a phytosaurian reptile from the Trias of Fort Lee, North America, as found in the rock. Phytosaurs, though not closely related to crocodiles, were very crocodile-like in appearance and probably led a similar life; they are among the most abundant of fossil reptiles from the later part of the Triassic period.

To find out its weight it would also be necessary to know the specific gravity of a dinosaur. Naturally, this cannot be measured directly, but since the living crocodilians are closely allied to dinosaurs, Colbert thought that their specific gravity would give an acceptable value. He then measured a young alligator and arrived at the specific gravity 0·9. It was now easy to calculate the probable weights of various dinosaurs, and Colbert's values will be given when the types of dinosaurs are described.

Dr Colbert observes that a larger alligator might have a heavier bone structure, relative to the volume, so that the exact value might be somewhat higher than 0·9. But, he adds, it does not seem worth while to tussle with a ten-foot crocodilian just for the sake of a few decimals. To those who might aspire to touching up the figures, it might be mentioned that the largest of all living crocodiles, the delta crocodile of the India-Australia region, may attain a length of up to 30 feet. To find the specific gravity of one of these would indeed be an achievement.

## Triassic dinosaurs

Most of the Triassic saurischians may be classed in a suborder called Palaeopoda by Colbert. The name means something like 'old-fashioned leg' and alludes to the form and function of the pelvic girdle and hind leg, which were taken over without further change from the thecodonts. In all of the latter the ilium is fairly short and extends mainly backward from the hip-joint. This means that the muscles pulling the leg backward and so propelling the body forward are well developed, while those that bring the leg up forward are less efficient. This probably worked well enough for ordinary walking, but not for bipedal running. In fact it seems that only those palaeopods that retained a four-legged gait were successful in the long run.

In the later Trias there appeared a number of bipedal forms, which were evidently carnivorous; they are called palaeosaurs ('old reptiles'). Certain palaeosaurs were no larger than the thecodonts, while others grew to become great, powerful predators up to twenty feet in length.

The systematics of the Palaeopoda are not yet very clear, mostly because the finds are incomplete. Colbert divides them into two groups, the palaeosaurs on one hand and then the plateosaurs ('flat reptiles' – an example of a scientific name which it is really superfluous to translate). These were large, bipedal reptiles, resembling in many respects the mighty sauropods next to be described. Some plateosaurs appear to have been meat-eating, others were plant-eaters; they had a rather small head, a long neck, and a long tail; the hind legs were longer and more powerful than the arms. These animals, too, grew to a length of some twenty feet and must have weighed as much as a big rhinoceros.

At the end of the Trias the palaeopods died out. In the view of Colbert and many others the gigantic sauropods may be derived from one of the groups in this suborder, the plateosaurs. Other authors, however, disagree on this point.

## The giant dinosaurs

The sauropods (Sauropoda), which appeared at the beginning of the Jurassic, are thought by Colbert and others to be easily derived from the plateosaur type, if it is imagined that the plateosaur got down on all fours (because of its increasing weight) and the neck and tail got still longer. On the other hand A.J.Charig, J.Attridge and A.W.Crompton consider that the sauropods more probably evolved from four-footed ancestors without a bipedal intermediate stage of the plateosaur type.

The sauropods may well be regarded as the most fantastic and astonishing tetrapods that ever existed. To many people they are almost synonymous with the dinosaurs, and many complete skeletons of these creatures are known. Yet they are enigmatic in many respects.

The history of the sauropods is a long one. It starts in the early Jurassic and continues right up to the end of the Cretaceous –

a stretch of time of some 120 million years. This should make it quite clear that the Sauropoda were not a temporary, 'overspecialised' type, which otherwise would seem a tempting idea; for the closer they are studied, the more extravagant do they appear.

Externally they were all rather alike: a short, massive torso on tall, pillar-like legs; a small head on a long neck; a long, sometimes enormously elongated tail. The longest sauropods attained lengths of up to 90 feet.

On closer scrutiny they reveal some variations on the basic theme. *Brontosaurus* or the Thunder reptile from the late Jurassic Morrison Formation in America is a well known representative of the Sauropoda. It reached a length of some sixty or seventy feet, and weighed about 30 metric tons. It walked on sturdy legs with short, broad feet; the hind feet carried three large, curved claws, the hands only one.

*Diplodocus*, a contemporary of *Brontosaurus*, is one of the lengthiest sauropods, reaching $87\frac{1}{2}$ feet in one skeleton. It was however rather slender of build and is estimated to have weighed little more than 10 tons – about twice as much as a big modern elephant. The long, narrow neck carried a very small head, with slender jaws and some feeble, peg-shaped teeth in front. The tail formed an enormous whiplash, undoubtedly an efficient defensive weapon.

A much smaller form was *Camarasaurus*, which lived both in Europe and North America. It reached a length of 15 feet and had a relatively short neck and tail. It is known both from the Upper Jurassic and Lower Cretaceous.

In all these forms the fore legs are much shorter than the hind, so that the body is highest in the pelvic region and the back slopes towards the shoulders. This has often been regarded as an inheritance from two-legged ancestors, but is not necessarily so. Many thecodonts probably remained functionally four-legged even if their legs were longer than their arms.

An exception from the rule is the immense *Brachiosaurus* ('arm reptile'), in which the fore legs were longer than the hind. A skeleton of a brachiosaur from Africa has a total length of 75 feet

and a shoulder height of 20 feet. The great height of the fore legs and the length of the neck raises the head of the animal to a height of almost 40 feet. The nostrils are on a raised eminence on top of the head, so that *Brachiosaurus* could stand in water this depth and reach the surface to breathe.

The tail in *Brachiosaurus* was rather short, the body immensely large. Presumably, immense strength of muscle and bone was necessary to enable the animal to suck down air into its lungs against the great water pressure. The weight of this specimen in life has been estimated at the utterly fantastic value of 78 tons. Thus, this reptile reaches the same weight class as the biggest whales, and as far as we know is exceeded only by the blue whale, which may weigh over 100 tons. However, this skeleton does not represent the maximum size in *Brachiosaurus*. Isolated bones of the same species have been found, exceeding considerably the corresponding bones in the complete skeleton, and suggesting that

29 The carnivorous *Antrodemus* (length thirty feet), here shown bending over the carcass of the sauropod *Brontosaurus*, comes from the late Jurassic Morrison beds of Colorado; its forelegs are much smaller than in *Plateosaurus*, and the neck is shorter, but the head is much larger, and armed with sharply pointed teeth.

other individuals reached even greater dimensions. It is probably realistic to assume that the largest brachiosaurs weighed up to 100 tons. This animal lived in the Upper Jurassic, and its remains have been found both in Africa and Colorado.

The weight in *Brachiosaurus* seems quite extraordinary in a land animal. How could these colossi move about on land without their limbs giving way? Can bone and muscle tissue really withstand such stresses? It must be remembered that the strength of bones and muscles grows in proportion to their cross-section, that is to say as the second power of the length; but the weight increases in proportion to the third power of the length. For instance, if a reptile grows to twice its former length, its weight will be eight times as great as before, but its limbs will only become four times as strong. This means that body proportions have to change with increasing size: the thickness of bones and muscles has got to increase much faster than the length of the body. There are of course mechanical limits to what is possible in this respect. If it is further remembered that the actual sliding surfaces of the joints in reptiles are rather less efficiently built than in mammals, we may get some inkling of the mechanical problems introduced by the giant proportions of the sauropods.

The construction of the backbone gives clear evidence of the stresses brought about by the great weight. The vertebrae were very large, and were joined together by powerful ligaments; their own weight was kept at a minimum by a cavernous structure combining lightness and strength.

It must be assumed that the sauropods normally dwelt in water, where their weight would cause no trouble, since it was buoyed up by the water. That they were in fact largely aquatic is seen from the position of the nostrils on top of the head. If we look about for a modern counterpart of the sauropods, it is perhaps the hippopotamus that comes to mind. It is evident that the sauropods, like the hippopotami, had not entirely forsaken the ground. Their limbs and feet always retained the structure seen in land animals, and must have been used for walking on the ground. They almost

30 One of the mysteries in the natural history of the Sauropoda is their diet; the feeble, peg-shaped teeth of Diplodocus (skull, (a)) does not dispel it. Another problem is the mechanical one introduced by immense size and weight; how it was met is shown by neck vertebra of (b) *Brontosaurus* in which all unnecessary bony tissue has been eliminated.

certainly emerged from the water to lay their eggs. However, it seems reasonable to assume that such excursions were tiresome, and that the sauropods were pleased to return to the water.

How could the sauropods grow so large? Did they grow very rapidly like the modern big whales, or did it take many decades for a sauropod to attain adult size? We cannot answer these questions with any certainty, although the paucity of fossils of half-grown sauropods might indicate a rather rapid rate of growth at least in the earliest years. One character typical of giant animals is the great development of the pituitary body, a gland which produces a growth-stimulating hormone. This has also been noted in many dinosaurs.

Again, the sauropods appear enigmatic when you start to wonder what they lived on. A full-grown elephant consumes one or two hundred pounds of food a day and has to eat almost continuously about ten hours a day. It also has a highly efficient dentition. The sauropod was much bigger than an elephant but probably less active. Still, it must have needed large amounts of food. But when one contemplates the relatively tiny head with its feeble jaws and very modest dental battery, the problem of how they were able to satisfy their appetites appears almost insoluble.

They cannot be compared with the big whales. The great sperm whale is a carnivore, which hunts and devours the giant squid and other prey – and whatever the sauropods were, it is impossible to imagine them in the role of active carnivores. Then there are the baleen whales, which strain plankton in their immense mouths; but the sauropods had small mouths, and even if the dentition of *Brachiosaurus* looks a bit like a straining apparatus, this can hardly be said of *Diplodocus*.

The food, whatever it was, must necessarily have been very nutritious and at the same time very soft, perhaps like gruel. What it was is an unsolved problem. Usually some kind of water plant is suggested, but there are also less orthodox ideas. One author points out that a putrefying carcass would answer the need of concentrated and easily devoured nourishment. (On the other

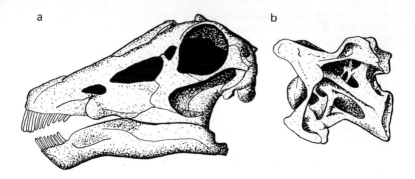

hand, probably the sauropods themselves were the only organisms producing carcasses of the size needed, and of course no group of animals can exist for 120 million years eating each other only, as in the village where everybody lived by taking in each other's laundry.)

The nervous system of the sauropods is also remarkable, but perhaps not as aberrant as has sometimes been stated. The small head of course accommodated only a very small brain, but this is a normal situation in all large reptiles and need not reflect on the nervous efficiency of the dinosaurs. Also, the brain was exceeded in size by a swelling of the spinal cord in the hip region. That is not so odd, because this was the pivot of the whole animal, with nerve fibres coming in from the hind legs to form a big plexus. It has been thought that the reflex passages might be correspondingly shortened, so that a touch of the tail would release a counterstroke even before the nerve impulse had travelled all the way to the head and back.

The idea of sauropods as big animals with feeble brains suggests a clumsy, sluggish creature. But a hippopotamus that moves heavily on land is transformed in water to a supple, graceful swimmer. The same may very well have been true for the sauropods, however heavily and slowly they may have stepped on land.

Finally, we have asked how the sauropods could grow so large, but we should also ask why: what was the advantage of their giant build? Probably their size and strength combined to safeguard them from attack from any but the very largest carnivorous dinosaurs. And that would be an important advantage, for apart from the whiplash tail they had no defensive weapons: no horns, no armour, no tusks. Incredible as this may seem, we have to admit that the giant size and whiplash tail made a successful combina-

31 Skeleton of *Diplodocus*, the longest sauropod known, from the late Jurassic Morrison beds, Utah. With its whiplash tail, this dinosaur attained a total length of more than eighty feet, but despite its length it was comparatively lightly built and probably weighed no more than two large modern elephants.

32 Life restoration of ostrich-dinosaur, *Struthiomimus*, one of a group of lightly-built, fast-running, medium-size forms (total length about ten feet). Present-day ostriches live on a varied diet including fruits, seeds, leaves, insects, and small vertebrates; perhaps the same was true for the struthiomimus, although egg-eating habits have also been suggested. The ostrich-like exterior does not necessarily imply a cast-iron stomach.

tion – no type of animal can exist for 120 million years without being excellently adapted to its environment.

In addition, the great size would tend to minimise heat-loss and enable them to retain a relatively high body temperature for some time after sundown.

Sauropods were common in Africa, Eurasia, and the Americas well into the Cretaceous. In northern North America they became extinct before the end of the Cretaceous, but in Texas they range up to the top of the Cretaceous, and likewise in Europe and Asia.

## Carnivorous dinosaurs

As early as the Keuper, side by side with the palaeopods, there appeared some dinosaurs (the *Theropoda*) with a more advanced type of pelvic girdle. Here the ilium has been extended forward along the backbone, giving a better leverage for the muscles that move the leg forward. These animals were thus capable of stepping more rapidly, and their speed and balance were greatly improved in comparison with the thecodonts and palaeopods.

The Triassic members of the suborder Theropoda are comparatively small and belong to the *Coelurosauria*, a group that survived to the end of the Cretaceous and thus was the longest lived of all dinosaur infraordinal groups. The coelurosaurs produced a series of small and medium-sized forms – the largest attained a length of about ten feet, but most were much smaller. They were active, dashing carnivores, bipedal but possessing well developed arms and hands, which may often have been used to help seize the prey.

A Triassic form in Europe, *Procompsognathus*, had a four-fingered hand; in the contemporary American *Coelophysis* there were only three fingers, and the same holds for the Jurassic forms like *Compsognathus* and *Ornitholestes*, known from Europe and North America respectively. The foot was four-toed, but the small first toe was at an angle to the other three; the structure is very bird-like. Coelurosaurs of this type are ubiquitous in all parts of the world, including Australia.

These carnivores probably hunted smaller lizards, mammals, insects and birds. The head was rather small and the jaws were not built to cope with large prey. As in birds, both vertebrae and long bones were hollow and apparently contained air sacs to reduce weight and increase speed.

In the later Cretaceous there appeared an offshoot of the coelurosaur group in the form of the ostrich-dinosaurs, for instance *Struthiomimus*. As in its relatives the hand was a three-fingered, clawed prehensile organ. The body was extremely slender, the legs very long and the small head was carried on a long neck. The resemblance to an ostrich was increased by the fact that the jaws were completely toothless and apparently armed with a horny bill as in a bird.

The mode of life of the ostrich dinosaurs has been the subject

33 Skeleton of *Antrodemus*, the great carnivorous dinosaur of the late Jurassic Morrison beds. Many skeletons of this form have been found at a highly productive site near Cleveland, Utah; now at an altitude of about 5,800 feet, it was

of much speculation. The combination of extreme speed, prehensile hands and weak, toothless jaws is indeed puzzling. It has been suggested that they lived on dinosaur eggs: the hands to seize the eggs, the speed to avoid the enraged parent, and the bill to break the shell – no teeth are needed to eat a raw egg. Although this idea may be rooted in a somewhat exaggerated view of dinosaurian family life (a topic that will be treated later on), this may sound fairly plausible, and there is actually a find from Mongolia that may support the theory. Here the crushed skull of an ostrich dinosaur was found on top of a nest with the eggs of another species, as if the egg-robber had been caught red-handed.

The suborder Theropoda comprises, besides the coelurosaurs, another group which is even more remarkable and indeed somewhat nightmarish. These are the great carnosaurs ('meat reptiles'), possibly evolved from Triassic coelurosaurs. They appeared in the

in Morrison times a low-lying, swampy area. This skeleton of
more than 425 bones was assembled from the incomplete skeletons of
three average-sized adult individuals, to make a superb mount
in a very life-like position. The length is thirty feet nine inches;
more than one-half is the long tail, which counterbalanced
the powerful, massive body.

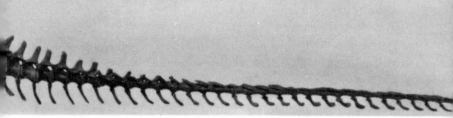

shape of tremendous predators in the early Jurassic.

*Antrodemus* may be described as a typical representative of these carnivorous dinosaurs, the megalosaurs and allosaurs, which flourished from the beginning of the Jurassic up to mid-Cretaceous times. The body length of *Antrodemus* measured up to thirty feet from the nose to the tip of the tail; more than half being the great tail, which functioned as a balance. The body was rather short and massive, and the neck was, in relation, shorter than in the coelurosaurs. The head was very big with enormous jaws. In order to reduce the weight of the skull, the cranial walls were fenestrated, so that the bony framework of the head looks like basket-work. The jaws, armed with big, sharp, recurved teeth, were somewhat elastic – the lower jaw was hinged in the middle – and this enabled the animal to bolt great chunks of flesh. The arms were short but powerful enough, and the three fingers carried long claws, so it

would seem that the hands were used together with the jaws to seize the prey. The hind legs were strong and the foot like that of a bird, with three big toes pointing forward and the small first toe to the rear.

This terrible creature has been estimated to weigh about two tons. It certainly was an active predator, which in spite of its great weight moved with long strides; it probably also scavenged on carcasses. It probably sought its prey mostly among the medium-sized herbivores, for instance the ornithischian dinosaurs. It has also been assumed that the allosaurs preyed on the largest of all the dinosaurs, the sauropods. But *Antrodemus*, though a monster to us, was almost a dwarf beside these giants. *Diplodocus* was one of the most lightly built sauropods, yet it weighed five times as much as an allosaur. For *Brontosaurus* the corresponding weight factor was 13, for *Brachiosaurus* more than 35. Is it possible that *Antrodemus* would assault creatures so much larger than itself?

A remarkable series of tracks has been found on a slab of hardened mud from the early Cretaceous; it is now displayed in

34 The small boy sitting in a sauropod footprint well demonstrates its huge size. To the left are tracks of carnivorous dinosaurs that stalked the sauropod. Trackways of dinosaurs on ancient mudflats, sealed in and preserved by rapid burial under later sediments, may give information on the stance, mode of walking, and even (as in this case) the behaviour of these long extinct animals. These footprints, from the early Cretaceous Trinity beds of Texas, are now mounted with a *Brontosaurus* skeleton.

the American Museum of Natural History. Here are the tracks of a gigantic sauropod, perhaps *Brontosaurus*, which walked here while the mud was still soft. Each footprint is more than five inches deep and could be used as a baby bath. But here are also the prints of large, bird-like, clawed feet – the tracks of an animal like *Antrodemus*. And it is quite clear that the carnosaur has been stalking the giant sauropod; it even stepped into the footprints of the other animal.

On the other hand it is known that present-day large carnivores, like lions, bears and the like, generally attack only animals in the same size class as themselves, or smaller. A lion will avoid an elephant, and a bear leaves the bison in peace. The prey, in other words, has to be manageable to the predator. An exception is formed by animals that hunt in packs, for instance the wolves, which may then overpower much larger animals. Did *Antrodemus* hunt in packs? Had the allosaurs evolved the complex set of instincts necessary for such behaviour? We are driven farther and farther into the realm of speculation.

*Antrodemus* was one of the biggest carnosaurs of its time. Other, related forms are smaller, down to about ten feet in length. A highly remarkable form is the fifteen-foot *Ceratosaurus* ('horned reptile') and its relatives, which generally resembled allosaurs but possessed five-fingered hands and a nose horn! A horned carnivore is a rather sensational kind of beast; otherwise such defensive weapons are almost the monopoly of plant-eaters. But among carnosaurs this character is found, not only in the ceratosaurs but also in at least one species of *Megalosaurus* from England.

Towards the end of the Cretaceous all the megalosaurs and most of the allosaurs became extinct; a related, surviving group are the peculiar spinosaurs, found for instance in Egypt. In these the spines of the backbone grew very long, up to six feet, and evidently supported a skin fold shaped like a fin along the back, resembling that of some pelycosaurs in the early Permian. Its function, which is unknown, was probably the same in both groups. It is just possible that it was used in demonstrative behaviour of some

35 *Left* The extraordinary carnosaur *Ceratosaurus*, which carried a horn on its nose – a defensive structure generally found only in herbivores. Although it did not reach the giant size of its contemporary *Antrodemus*, also from the late Jurassic Morrison beds, it was nevertheless an awe-inspiring twenty feet.
*Below left* Skeleton of *Tyrannosaurus*, greatest of all carnivorous dinosaurs. The length is about fifty feet, and it stood eighteen feet to the top of its head. The small forelegs appear useless, yet are not degenerate and probably had some function. A resting tyrannosaur lay flat on its belly; when rising, the fingers may have hooked into the ground to keep the body still as the great hind legs straightened.

*Below* The light, almost lacy build of the great *Ceratosaurus* skull is typical of the allosaurs and megalosaurs.

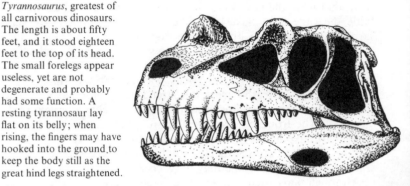

kind – this type of behaviour is quite common in birds. But it is perhaps more probable that the sail functioned as a heat-regulator, in the way that A. S. Romer has argued for the pelycosaurs.

In a hot climate the muscle work during a swift rush may raise the body temperature of a reptile to such a pitch that the animal is threatened by heat-stroke and has to keep still and cool off for a long while. A fin or sail with a rich web of blood vessels might then be an efficient device for dissipating the surplus heat. On the other hand the reptile will be sluggish when its body temperature is low, for instance after a cold night. By then turning the fin to catch the sun, the spinosaur would be able to warm up more rapidly, and become active.

The most important carnosaurs in the late Cretaceous were however the enormous deinodonts, with forms like *Deinodon* ('terror tooth'), *Gorgosaurus* ('terrible reptile') and *Tyrannosaurus*. Among these the tyrant reptile was the largest and also the last to appear. With its length of 47 feet and an estimated weight of almost seven tons it was the biggest terrestrial carnivore that has ever existed, as far as we can tell.

The great head of *Tyrannosaurus* attains a length of more than

36 Sculpture restoration of *Hypsilophodon*, a small ornithopod dinosaur from the lower Cretaceous of southern England (length up to five or six feet). It is a primitive form that survived alongside more advanced dinosaurs; it may have been tree-living. If this is correct, it may be compared to the opossum, a modern example of a primitive form surviving in an arboreal environment.

four feet. It is more massively built than that of *Antrodemus*, but as in that form the jaws were elastic to some degree; when opened wide the gape must have been like a cavern. The teeth are very large and sabre-like, with sharp serrated edges.

Many restorations and mounts show *Tyrannosaurus* and other deinodonts in a very upright position, which makes them about twenty feet tall. However, the body was probably normally held more inclined, with the great tail stretched out for balance. The hind legs are very long and powerful with the characteristic bird-like foot. The front legs, conversely, are quite stunted and do not appear to have had any function at all; they were even too short for the animal to scratch its chin. Only two fingers were present in *Gorgosaurus*, and the same may have been true for *Tyrannosaurus*.

Although most deinodonts were large, there are also smaller forms in the family. One of the smallest had a skull only eight inches long, less than a sixth of the length in *Tyrannosaurus*. This little deinodont probably weighed only some 70-80 pounds, so that it was about the size of a large dog.

The diet of the big deinodonts appears less of a problem than in the case of the allosaurs, for in the late Cretaceous there existed a great number of ornithischians of suitable size. Most of them were somewhat smaller and lighter than *Tyrannosaurus*.

## Bipedal ornithischians

The ornithischians never grew to the size of the big sauropods; neither did they produce as highly specialised bipedal forms as the carnosaurs. On the other hand they evolved in much greater diversity than the saurischians, which were restricted to two basic types.

All the ornithischians seem to have been herbivorous, and their teeth are more complicated in structure than those of the saurischians. In most forms the jaws are toothless in front, and there probably was a horny beak. Like the saurischians, ornithischians evolved from thecodonts tending to a bipedal gait. Ornithischians

occur as early as in the late Trias of South Africa, but finds are rare and not yet fully studied.

The bipedal ornithischians constitute the suborder Ornithopoda. They generally moved bipedally, but the arms of all the ornithopods were well developed and functional; they were evidently used for temporary support and when climbing or swimming, as prehensile organs and, in some instances, for defence.

A group of small ornithopods, still rather close to the thecodonts, is represented by *Hypsilophodon* (a long name alluding to the shape of the teeth with their high enamel folds). In this form, which is known from the later Jurassic and the Cretaceous, there are teeth in the front of the jaws, which is unusual in ornithischians. *Hypsilophodon*, of which several skeletons have been excavated in the Wealden beds, grew to a length of five feet. The hand carried five fingers, but the little finger was reduced and protruded at right

37 Skeleton (*below*) and restoration (*left*) of the herbivorous ornithopod dinosaur *Camptosaurus* from the late Jurassic Morrison beds of Wyoming. Length in this genus varied from four to fifteen feet. Apparently an inoffensive browser, with flight as its only resort when attacked, *Camptosaurus* had a more upright posture than the carnivorous dinosaurs. The toothless, beak-like front end of the jaws may have been used to pluck off leafy twigs that were then chewed by the cheek teeth.

38 *Iguanodon*, the large ornithopod with the spur-shaped thumb, is one of the most abundant dinosaurs in the early Cretaceous. Among the species known, *Iguanodon bernissartensis* from Belgium is the largest (length about twenty-five feet); shown opposite is an isolated skeleton; the frontispiece shows a herd of adult individuals which met a simultaneous death, perhaps from falling into a crevasse. The proportions of the body and tail indicate a very upright posture, and a large herd of these tall animals, reaching heights of twelve to fifteen feet, must have been a very impressive sight.

angles from the wrist. The foot was four-toed, and all the toes pointed forwards in contrast with the condition in the theropods.

The dental battery suggests that *Hypsilophodon* was herbivorous, and a certain resemblance to the present-day tree kangaroos has led to the assumption that it was a tree-living form, which foraged for fruits and leaves in the foliage. The long and strong toes and fingers do seem well adapted for a climbing way of life. On the other hand *Hypsilophodon* has a double row of small bony plates along its backbone, hardly an arrangement you would expect to find in an arboreal form. However, this may represent a heritage from the thecodonts, in which such plates were common.

*Camptosaurus* from Europe and North America grew to much larger size than *Hypsilophodon* but does not belong to the largest ornithopods; it attained a length of up to 17 feet, and its weight has been estimated at some 1600 pounds. It is a member of the iguanodont family. In this typical ornithopod no really specialised characters are yet to be seen, except that teeth are lacking in the front of the jaws. The cheek teeth were well developed and suggest that *Camptosaurus* was a browser. The arms are comparatively long and were probably used for support in a squatting position, and perhaps also to seize twigs and leaves and carry them to the mouth. The claws of the five fingers were flat and hooflike. The tail was powerful but shorter than in the theropods, and hence a less efficient counterbalance; so that the normal poise of the body in the ornithopods was probably more upright than in the theropods, thus reducing the torsional force acting at the hip pivot.

A close relative of *Camptosaurus* may be seen in *Iguanodon*, which is known from the Wealden in Europe and was the first dinosaur to be discovered. Two or three species are known. The largest, *Iguanodon bernissartensis*, of which seventeen skeletons were found in a coal-mine at Bernissart, Belgium, reached a length of almost 25 feet and may have been about twelve feet tall when

sitting up. The great number of skeletons no doubt comes from a herd of these animals, which fell down a precipice, perhaps in fleeing from a pursuing carnosaur. This gives us the interesting information that iguanodonts were social and moved in herds. It is also interesting to note that all of the specimens are adult: there is no mixing of different age groups, and so the young animals did not herd with their parents. Many footprints, apparently from the three-toed hind feet of *Iguanodon*, have been found in the Wealden.

*Iguanodon* was built on the same lines as *Camptosaurus*, but its hand is quite unique because the thumb has been converted into a strong spike or spur, protruding at right angles to the other fingers. The first isolated finds of this spike were interpreted as horn cores, so that old restorations of *Iguanodon* showed it carrying a horn on its nose!

The spur is probably to be interpreted as a defensive weapon, for it is found in both sexes and thus cannot be a secondary sexual character like the spur of a cock. Perhaps *Iguanodon* had evolved the instinct to strike at the eyes of its adversary, the only point where such a weapon can have had any important effect on an allosaur or megalosaur.

The earliest well-known ornithopod, *Heterodontosaurus* from the Trias of South Africa, had cheek teeth not unlike those of *Iguanodon*, apparently suggesting leaf-eating habits, but also possessed big tusk-like eye-teeth almost like those of a carnivore. They were probably defensive structures. It was a small animal with a head only about four inches long.

In certain Upper Cretaceous ornithopods we may note a seemingly most bizarre specialisation. On first sight the creature called *Pachycephalosaurus* has a quite intellectual look about it, with its high forehead and great domed cranium. The impression is however entirely misleading, for the head consists mostly of solid bone, in the centre of which is the small brain. The name, meaning thick-headed reptile, is thus very apt.

The length of these bonehead dinosaurs was up to six or seven feet, so that they are among the smaller ornithopods. They were bipedal with strong, four-toed hind legs and short arms. The head was decorated with numerous knobs of different sizes, above which the domed crown of the head arose, smooth as a human bald pate.

The reason why they developed such an immensely strong armour around the rather primitive brain is obscure. They do not seem to have had any other kind of armour.

## Duckbilled dinosaurs

A new, important group of ornithopod dinosaurs appeared in the Upper Cretaceous: these are the duckbills, or family Hadrosauridae. ('*Hadrosaurus*' means simply 'big reptile'.) A great number of skeletons of these dinosaurs has been found in the Upper Cretaceous of North America, but they are also known in Asia and

39 Skeleton of the crestless duckbill *Anatosaurus* in walking pose (for a partly mummified specimen in the death pose, see p. 14). This form, which reached a length of thirty to forty feet, was found in the uppermost Cretaceous Lance beds of North America. Note the flattened, toothless 'duckbill', the battery of cheek teeth, the powerful arms, and the compressed swimming tail.

Europe. The body length varied about an average of some thirty feet, and the structure was in general not unlike that of *Iguanodon*. As in that form, the foot was three-toed and the arms comparatively powerful; but the little finger had been lost, the thumb was reduced, and the three remaining fingers carried small hoofs. The weight of *Hadrosaurus* is estimated at about three tons, while the somewhat larger *Corythosaurus* may have weighed 3·8 tons.

Hadrosaur 'mummies' have also been discovered, or specimens in which parts of the skin and other soft parts have been preserved in dried condition and then fossilised. These show that the skin was covered by a mosaic of small scales. The fingers supported a skin fold, and apparently the foot was webbed too. The tail was laterally compressed as in a crocodile and was probably used in the same way. It is evident that the duckbills were powerful swimmers and that they spent part of their life in water.

It is the shape of the head and jaws that has suggested the name duckbills. The jaws were broad and flat in front and probably were covered by a duck-like horny bill. The dentition is however impressive, whereas ducks are of course toothless. The teeth of the duckbills were simple leaf-like structures as is normal in ornithischians, but here their number had increased enormously, and they were tightly packed, up to six hundred in each jaw half. This means that a single individual could have more than two thousand teeth in function at a time. The tooth batteries formed an extremely efficient grinding apparatus, which also had the great advantage that it could not be worn out; for worn teeth were continuously replaced by new. It may thus be said that the hadrosaurs were even superior to such efficient herbivores as cows and elephants, for in these the longevity is always dependent on the wear of the teeth.

The duck-like bill has led many students to assume that the hadrosaurs led a life analogous to that of modern ducks. It used to be thought that they inhabited shallow waters, where they rooted in the mud for plants, crustaceans and the like. But, as J. H. Ostrom has remarked, what is true for a duck need not necessarily be true for a thirty-foot dinosaur weighing several tons. Further, the tremendous dental battery of the hadrosaurs is completely unlike the toothless hind part of a duck's bill.

Fortunately the diet of the hadrosaurs is known from a study of the stomach contents of one of the 'mummies', a specimen in the Senckenberg Museum in Frankfurt. This duckbill menu was published as early as 1922 by R. Kräusel. He found a mass of needles of a coniferous tree (*Cunninghamites*) and twigs, seeds and fruits from other land plants. No remains of water plants or water animals were found. So it must be concluded that the

40 Sections through head of two crested hadrosaurs,
(a) *Parasaurolophus* and (b) *Corythosaurus*, to show course of nasal
tubes inside crests. From the point where left and right nasal tubes joined
on top of the head, a single tube (not shown) passed straight down
to the windpipe and lungs. Convolutions of nasal passages are now
thought to have augmented the sense of smell in hadrosaurs.

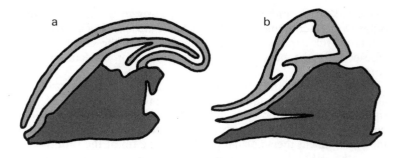

hadrosaurs sought their food on land, in the woods, and that they actually used their splendid dentition to grind conifer needles and other abrasive foods.

Still it is clear enough that the hadrosaurs show swimming adaptations and probably spent part of their life in the water. Ostrom suggests the following solution to the seeming contradiction. The hadrosaurs, normally terrestrial browsers, lack any kind of defensive armament against the contemporary carnosaurs; they did not even have the thumb spur of an *Iguanodon*. Their safety lay in flight, but though they probably were fairly fleet of foot, it is unlikely that they would have been able to outrun the carnosaurs. Their best way of escape would have been to jump into the water and swim off, for it is clear that they were better swimmers than the carnosaurs.

This line of thought is supplemented in an interesting way by a specialisation quite unique to the hadrosaurs, and recently re-studied by Ostrom. It is the evolution of a peculiar bony crest on top of the head, which may assume very bizarre shapes. In *Hadrosaurus* itself the crest is lacking. In *Kritosaurus* it is barely present looking somewhat like a Roman nose (hence the name, meaning 'noble reptile'). *Corythosaurus* or the helmeted reptile had a crest shaped like a Corinthian helmet; in *Lambeosaurus* it is a grotesque outgrowth resembling a top hat; while *Parasaurolophus* carried a recurved, tubular crest. Dissections show that the crests in each case contain the nasal tubes, forming more or less complicated convolutions.

These chambers and convolutions have generally been considered as an adaptation for aquatic life. They have been compared with snorkels, making it possible to breathe in the submerged position, but this is manifestly incorrect, for the nostrils are not on top of the crests but in their normal position at the tip of the snout.

On the other hand the crests have been regarded as storage chambers for air, thus making it possible for the hadrosaur to make longer dives. But this can also be shown to be impossible. In the first place, the amount of air accommodated by the crests would be only a few per cent of the lung capacity, so that it would really be almost useless. Besides and more important, there is no possible way of getting the air from the crest down into the lungs in the submerged position, except by inhaling water at the same time, which might result in drowning.

A more reasonable suggestion, made by C. Wiman, is that the crests may have functioned as resonating chambers. It is indeed probable that they did, and that hadrosaurs were as noisy as crocodiles, but it can hardly be the only explanation.

It appears most probable that the air passages in the crests were developed to serve the sense of smell. The olfactory sense cells are located in the epithelium on the inside of the nasal tube. In most reptiles the surface covered by olfactory epithelium is comparatively small. In the mammals its surface has been greatly increased, partly by a widening of the nasal cavity, and partly by the presence of thin, epithelium-clad bones within the cavity, the turbinals. Dinosaurs have no turbinals, and if they are dependent on the sense of smell, the only way to improve it seems to be to lengthen the nasal passage itself, as has happened in the Duckbills.

We can now picture to ourselves these duck-billed dinosaurs as they were in life. They probably kept fairly close to streams and lakes. They browsed on leaves, conifer needles and twigs, which were seized in the flat bill, perhaps with the aid of the hands. As soon as they scented danger, however, they would rush headlong for the shore, perhaps signalling to each other by crying out; there would be a series of great splashes as the big creatures hurled themselves into the water, to swim to some other, safer shore, or to hide in one of the numerous river channels in the great delta.

Many entire skeletons of duckbilled dinosaurs have been excavated in such a swimming position, but with the head thrown back, as if in death-throes. Are these remains of animals that were attacked by carnosaurs, received mortal wounds but broke away and struck out into the water, before they finally succumbed?

## Stegosaurs

While the main deployment of the saurischians took place in the Trias and Jurassic, so that the Cretaceous forms are generally but variations on a long-established theme, the culmination in ornithischian evolution occurred in Cretaceous time. From the Jurassic only two ornithischian suborders are known: the Ornithopoda, which have already been described, and the Stegosauria.

The stegosaurs (Stegosauria) initiate the series of four-footed ornithischians. They appear in the Lias with an incompletely known form, *Scelidosaurus*, which attained a length of about twelve feet. It is known that it possessed bony plates somewhat resembling those of later stegosaurs, but the exact arrangement is not known.

The best-known form is *Stegosaurus* from the late Jurassic Morrison Formation; it is also found in Europe. This was a fairly large, heavily built dinosaur up to twenty feet in length. The neck and tail were rather short, so that most of the length is in the body. The heritage from bipedal ancestors may be seen in the fact that the hind legs are much longer than the arms, so that the back slopes very markedly forward, and the small head is close to the ground. The foot is three-toed, but the hand carries five fingers.

41 Life-size model of the 'plated dinosaur' *Stegosaurus* from the late Jurassic Morrison beds in North America. Length of original twenty feet, height at hip plate eighteen feet. Though some smaller, related forms may have been bipedal, *Stegosaurus* with its great weight and clumsy armour was clearly down on all fours for good.

Both fingers and toes terminate in hoof-like nails. The weight of *Stegosaurus* has been estimated by Colbert at 1·8 tons.

The oddest thing about *Stegosaurus* is the adornment or armour formed by two rows of large, triangular bony plates along the neck and back, and part of the tail: They apparently alternated in the manner shown in the restoration. The plates over the hip were the largest. Near the end of the tail the plates are succeeded by big spikes, of which there are two pairs.

No other kind of armour is present, so that *Stegosaurus* was nearly defenceless from the flank. Perhaps it resisted attack by curling up like a hedgehog, head turned away from the attacker, while the plates were spread out and the tail lashed sideways.

The small, elongate head of *Stegosaurus* carried a rather feeble dentition consisting of about twenty teeth in each jaw half; the front of the jaw was toothless. This animal presumably lived on some kind of low-growing plant. The problem is rather similar to that of the sauropods. Here again we see a large dinosaur getting along very well with a poor dentition and a very small head.

*Stegosaurus* has attained some undeserved fame on the score of smallness of brain; the true brain was greatly exceeded in size by the swelling of the spinal cord in the hip region. There is however no reason to refer to this as a 'second brain' or to assume, even in a metaphorical way, that the stegosaur was able to 'reason *A priori* as well as *A posteriori* . . .' as claimed by one of the American minor poets. Actually, it must be repeated, *any* large reptile will be found to have a comparatively small brain. This is true, for instance, of a large crocodile or python, and there are few who would regard these as dull and harmless beasts! The large size of the posterior 'brain' is shown by Tilly Edinger to be dependent on the large size of the parts innervated from here. It may be noted that the living ostrich also has a corresponding swelling larger than the real brain.

A Upper Jurassic East African form appears to have only a few plates on its back, while there were spikes on the tail and the front part of the body. The stegosaurs persisted in the Lower Cretaceous, then became extinct; they are thus the only ornith-

ischian suborder that did not survive until the end of the Cretaceous. Our judgment is no doubt coloured by the fact that we have never seen a living dinosaur, but we should perhaps admit that the stegosaurs look a less successful model than the others, so that their early demise is perhaps not to be wondered at. Still, they did survive for some fifty million years.

In popular accounts you may happen upon a restoration showing *Tyrannosaurus* helping itself to a stegosaur steak, but that is as anachronistic as the cartoons showing cave men and dinosaurs together. When the tyrant dinosaur lived, there was nothing left of *Stegosaurus* but fossil bones many million years old.

### Armoured dinosaurs

In Cretaceous times the stegosaurs were succeeded by Ankylosauria (the name really means 'crooked lizard' and alludes to the strongly curved ribs). They were heavily armoured forms not unlike modern armadillos. The ankylosaurs appeared at the beginning of the Cretaceous and thus existed side by side with the stegosaurs for some time.

In *Polacanthus*, a contemporary of the last stegosaurs in Europe, there is an arrangement bringing *Stegosaurus* to mind. It is a double row of big spikes along the back and tail; the hip region was protected by a coat of mail, consisting of a mosaic of small bones.

Late Cretaceous forms like *Ankylosaurus*, *Palaeoscincus* and

42 One of the earlier (Belly River stage) ceratopsian dinosaurs, the single-horned *Monoclonius*; length about sixteen feet. Note the bulge formed by the temporal muscle playing over the neck 'frill'. Most restorations incorrectly show the frill as a bald, even fenestrated plate of bone. The curve of the horn shows that *Monoclonius* tried to impale its enemies simply by rushing at them; the reduced mobility of the enormous head made tossing impossible. Figure adapted from Lull.

*Nodosaurus* carried a much heavier mail. The entire back was encased in bony plates, and on the sides there were big spikes in a row. In some forms the fore legs were protected by particularly large spikes in the shoulder region, while the tail was entirely encased by a series of bony rings, and armed with long barbs. The small head was also protected by a thick roof. The body was broad and flat, and waddled on short legs; these dinosaurs have been likened to living tanks. The length of the body could be up to 25 feet, the shoulder height six feet; for *Palaeoscincus* a weight of $3\frac{1}{2}$ tons has been estimated.

The mode of life of the ankylosaurs may be thought to have been similar to that of the stegosaurs, whatever that was; for they have a similar, very small head with weak teeth – some of them were in fact completely toothless ('Nodosaurus' means 'toothless reptile'). Many other forms belong to this quite flourishing group, but most finds are too incomplete to permit accurate restorations.

## Horned dinosaurs

The *Ceratopsia* were the last dinosaur suborder to evolve; they appear only in the later Cretaceous. Their origin may probably be sought in a group of ornithopods, the psittacosaurs, which inhabited Eurasia in the earlier Cretaceous. *Psittacosaurus* means 'parrot-reptile'. In contrast with other ornithopods it had a comparatively high and narrow head with a parrot-like beak, which is retained by the true ceratopsians. Otherwise this three-foot dinosaur had the body of an ornithipod, that is to say it was bipedal; but the pelvic girdle is strikingly similar to that in the Ceratopsia, and the same is true of the hand (which is five-fingered) and the four-toed foot.

The most primitive true ceratopsian known is *Protoceratops*, which was discovered in the Upper Cretaceous Djadochta formation of Mongolia. Of this ornithischian no less than 80 skeletons have been found, showing all the stages in development from newly hatched young to adults measuring 5-7 feet; the adults

probably weighed about 900 pounds. In addition, there are great numbers of eggs. It is most unusual to find the juvenile stages so well represented as in this material.

*Protoceratops* had abandoned the bipedal pose and was down on all fours, probably because of the great weight of its big head. The head made up one-third of the total length of the dinosaur, and was as long as the body. The jaws formed a strong, curved parrot-beak in front, toothless except for four vestigial teeth in the upper jaw – mementoes of a fully toothed ancestor. The cheek teeth, which were present in great numbers, formed a shearing mechanism, as in other ceratopsians. At the back the skull flared out into a long, fenestrated collar or neck shield, which was absent in the newly hatched and grew out gradually during the life of the individual. *Protoceratops* did not have a horn with a bony core, but there was a rough area on top of the snout that may have served as a base for a small keratin horn in some individuals, as in a rhinoceros.

At a somewhat later stage in the Upper Cretaceous, *Protoceratops* gave rise to a mighty tribe of ceratopsians, which were richly varied in structure. Several evolutionary lines may be traced through the sequence Belly River – Edmonton – Lance. One of these lines starts out with *Monoclonius* from the Belly River stage. This was a dinosaur about sixteen feet long, with a single, long nasal horn rather like that in a rhinoceros, and a short, fenestrated neck shield; the name means 'one-stemmed' in allusion to the single horn.

At the Lance stage this line culminated in the monumental *Triceratops*, which attained a maximal length of twenty feet or more and weighed $8\frac{1}{2}$ tons. It was the heaviest ornithischian known to us, and was exceeded among dinosaurs only by the sauropods. *Triceratops* carried two big frontal horns; they curved outward and forward as in some bovines, and the long, pointed horn-core was evidently encased in a keratin sheath. The first find of *Triceratops* was a horn-core, which was then thought to belong to a fossil bison. *Triceratops* also had a nasal horn corresponding to that in *Monoclonius*, but it was shorter, and in some specimens the core is almost absent, although probably these individuals did carry a keratin horn.

The neck shield in *Triceratops* is a solid, saddle-shaped plate without fenestration, but with a jagged border. The body was very massive, the legs tall and pillar-shaped, and the tail powerful but comparatively short.

In another evolutionary line, which became extinct as early as the Belly River stage, the spikes along the borders of the neck frill developed into long spines; it is represented by *Styracosaurus* ('spiny reptile') which was about the same size as the contemporary *Monoclonius* and, like it, had a big nasal horn. It weighed about 3·7 tons.

A third line of evolution led to the appearance, in the Edmonton stage, of *Pentaceratops* with a very large, fenestrated neck frill. The armament of horns resembled that of *Triceratops*, but there was an additional pair of horns on each side of the frill. This animal was thus five-horned, as suggested by the name. In the Lance it was succeeded by the gigantic *Torosaurus*, which was almost as large as *Triceratops* but possessed a very long, fenestrated neck shield of the same type as in *Pentaceratops*.

The interpretation of the horns in these creatures is no problem, especially when one thinks of the contemporary deinodonts. They no doubt formed a defensive armament, and since males and females carried similar horns we may conclude that they were used against other species rather than in mating fights. There are

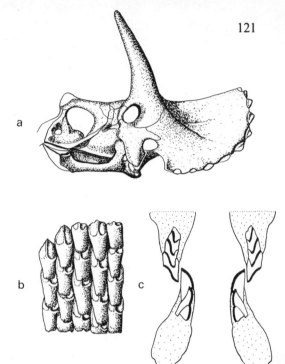

43 *Above* Skull and teeth of *Triceratops*. (a) Skull and mandible from Colorado, after Marsh. (b) Part of dental battery showing replacement cheek teeth lying in readiness below those currently in function. (c) Cross section of the upper and lower jaws, showing the scissor-like function when the lower cheek teeth glide in between the upper. After Ostrom. *Below* Skeleton of *Triceratops*, from the uppermost Cretaceous Lance beds of Wyoming; length twenty-four feet. It marks the culmination of ceratopsian evolution. Note the forward orientation of the horns, the toothless parrot-like beak, and the numerous cheek teeth, the bony neck frill, and the enormously heavy build.

examples of broken and healed horns, showing that they were indeed used. The structure of the front and hind legs indicate that the ceratopsians, when an enemy came into view, were able to swing round rapidly to meet the danger; they would move the rear about a pivot in the shoulder region. Probably a *Triceratops* of more than eight tons was a formidable adversary even to *Tyrannosaurus*. It is to be assumed that the defensive strength of these dinosaurs was an important factor in their almost explosive evolution and success in the Late Cretaceous.

Another typical ceratopsian characteristic is the neck frill. In *Triceratops* the head reached a length of eight feet, more than a third of the total length of the animal, and the big plate of bone accounts for about one half of this. In other forms the frill was still larger, especially in the *Torosaurus-Pentaceratops* group, where it may form two-thirds of the total length of the head.

The function of the frill was mainly connected with the development of the muscles closing the jaws, as shown by J. H. Ostrom. In addition it probably formed an important area of insertion for the muscles extending from the head to the neck and body. Such a big and heavy head had to be held and manoeuvred by very powerful muscles. A third function, which had indeed been regarded as particularly vital, was to protect the neck against attack, but this has probably been exaggerated. The horns would have made attack at this point difficult in any case, and the animal would have been better served by armoured flanks.

It is clear that immense masticating muscles operated the jaws of *Triceratops*. We must now turn to a scrutiny of its remarkable jaw mechanism. The cheek teeth form great tooth magazines. In each jaw half there were up to thirty-five tightly packed vertical columns of teeth, each consisting of a long series of teeth on top of each other, so that a new one would come into play as soon as the outermost was worn down. There was, however, no chewing at all in the ordinary sense. On the contrary, the upper and lower teeth slid vertically past each other like scissor blades. These enormous shears were operated by muscles three or four feet long!

No living animal has a dental arrangement even remotely like that in *Triceratops*, so that there are no analogies to guide us as to the diet of these creatures. At any rate it is evident that it was completely different from that of the hadrosaurs, which had grinding cheek teeth. Such a dentition is characteristic of browsing and grazing forms, but shearing teeth are usually found in carnivores only.

The ceratopsians, however, do not look at all like carnivores. The most likely suggestion would seem to be that they fed on some highly fibrous vegetable matter, and J.H.Ostrom thinks that in the Upper Cretaceous flora, cycad and palm fronds may have served as staple food; both were common at the time. Some of these plants were comparatively low-growing with a short trunk, and would have been within easy reach of the ceratopsians. Some of the taller stems could have easily been trodden down by these heavy beasts, enabling them to reach the fronds. Perhaps the parrot beak was used to pluck out the fronds, which were then pushed back in the mouth and cut up by the cheek teeth; the animal probably had muscular cheeks to keep the food from falling out.

### Dinosaur eggs

Most reptiles reproduce by laying eggs (ovipary). In a few forms, termed ovoviviparous, the eggs are hatched within the body of the mother; in other, rare instances we know of vivipary, or reptiles bearing young. In such instances a kind of placenta may be formed, transmitting nutriment and oxygen from the mother.

All the crocodiles are oviparous, and the same was probably true for most or all dinosaurs. Crocodile eggs may be deposited in a special nest, or the crocodile may bury its eggs in mouldering plant remains. In most cases, however, the eggs are buried in dry sand above the water table, where the temperature is constant and, thanks to the sun's rays, comparatively high. This situation may be conducive to fossilisation: if the eggs do not hatch they may remain in the sand and finally become preserved as fossils.

44 Dinosaur eggs. *Right* Eggs referred to the primitive late Cretaceous ceratopsian *Protoceratops*, found in the Gobi Desert, Mongolia; these are the first dinosaur eggs to be definitely identified as such.
*Left* Large dinosaur egg from site near Montpellier, southern France; this specimen may be near the maximum size attainable in a reptilian egg. Part of the shell has been lost, exposing the internal cast.

As early as 1869 egg-shells from Cretaceous deposits in France were described, but at that time it could not be decided whether they belonged to birds or reptiles. In the 1920's the American expedition to Mongolia discovered great numbers of eggs together with fossils of *Protoceratops*, which caused, deservedly, a great deal of sensation and publicity. Eggs or egg-shells have since been found in various areas, both in Europe and North America. In France there are particularly numerous eggs from various sites, but it is only in the last decade that these have been definitely shown to belong to dinosaurs, not birds as was initially thought.

The Mongolian eggs, which are ascribed to *Protoceratops*, had been buried in the sand. Some of them contain remains of embryos. The eggs had been deposited in circles, of which several were laid on top of each other; the total number of eggs in such a nest was 30-35. The eggs were oblong and about eight inches in length. There are also other types of eggs from Mongolia and from China.

In the Cretaceous deposits of Provence no less than nine different kinds of eggs have been differentiated in the course of studies by R. Dughi and F. Sirugue. The smallest have a content of 0·4 litres, while the largest average some 3·3 litres. That they are indeed reptile eggs has been shown by comparative studies of the shell structure in birds and reptiles, which show marked differences. In both cases the shell is built up by prismatic elements extending outwards from the inner surface of the shell, but the bird egg also has a very tough external layer, which is lacking in the reptile egg. That this was also the case in fossil bird eggs has been shown by study of Tertiary eggs of giant ground birds from the Paris basin.

Oddly enough, however, the Mongolian eggs are regarded by the French experts as bird-like. Yet they were discovered together with newly hatched and half-grown *Protoceratops* young, and it is difficult to imagine that they do not belong to this species. Perhaps a bird-like type of egg shell was evolved in the Ceratopsia.

How the nine egg types from Provence are distributed amongst various reptiles is hard to say, but the largest may well have belonged to sauropods. Two species of sauropods are known from fossil bones, *Hypselosaurus* – a relative of *Diplodocus* but smaller, measuring about thirty-five feet in length – and the considerably larger *Titanosaurus*. We also know of an ornithopod belonging to the iguanodont family, as well as sparse remains of an armoured dinosaur and a carnosaur (*Megalosaurus*). Only five species of dinosaurs are thus known here, but this is of course only a small fraction of the entire reptilian fauna of the time.

In spite of their enormous size, the sauropods did not lay particularly large eggs. They are not in the same size class as the famous elephant bird eggs from Madagascar, which can hold up to eight litres. This is probably close to the maximum possible. In a still larger egg the pressure of the internal fluid would be so great that the shell would have to be excessively thick, and as a result the young would find it difficult to get out. The reptilian egg shell is more fragile than that of the birds, so that the largest specimens from Provence may well represent their biological maximum.

At one of the Provençal sites you can still see how the dinosaur behaved when the eggs were laid. It deposited a few eggs beside each other, then took a step forward and repeated the procedure. It was repeated fifteen to twenty times, resulting in a corresponding number of clutches. The number of eggs in each group varies, depending perhaps on the species of dinosaur, from one to five.

In Montarnaud near Montpellier numerous dinosaur eggs are also found, but here they lie in long, winding rows, which may suggest that they have rolled out from their original position. These eggs are as round as balls. To see these tremendous eggs in the rocks, in the peaceful wooded countryside, is a memorable experience.

As with crocodiles, dinosaurs apparently did not practise any care of their young, which were left to fend for themselves after hatching. It follows that those restorations which show Papa and Mama *Triceratops* with their baby, keeping hungry *Tyrannosaurus* at bay, are completely unrealistic. The fossil *Iguanodon* herd demonstrates that the age groups did not mix. Although many of the herbivorous dinosaurs undoubtedly moved about in large herds, there was no family grouping.

We have reached the end of our survey of the dinosaurs. But dinosaurs were not the only creatures that populated our earth in the Jurassic and Cretaceous periods. Let us now go on to take a look at the rich fauna of other animals existing at the same time as the dinosaurs. Many of them are fully as remarkable and impressive as the dinosaurs themselves.

# er terrestrial vertebrates

The dinosaurs, as we have seen, were land animals or at most amphibious forms inhabiting freshwater basins and shallow coastal waters. The present chapter is devoted to a survey of those vertebrates that shared this habitat with the dinosaurs, while chapter 5 moves on to the marine environment alien to the dinosaurs.

## Crocodiles

If we begin with the allied orders of archosaurs, the crocodilians (Crocodilia) may be considered first. Now we meet with animals which can still be studied in the flesh – most of us have seen crocodiles, in motion pictures at any rate. In general the crocodilians may be described as four-footed animals, which however always reveal their thecodont ancestry in that the hind limbs are much longer than the arms, just as in the dinosaurs. When moving slowly the crocodilian may crawl with its belly on the ground, but it is also able to walk and even to run quite rapidly; it then raises itself high on its legs and looks not unlike a dinosaur.

Almost all the crocodilians are to some extent adapted to an aquatic life. In this connexion the head and body are broad and flat, while the tail tends to be laterally compressed and functions as a swimming organ. However, most crocodilians remain freshwater forms and spend much of their time out of the water, so that they can only in exceptional cases be classed among the fauna of the seas.

Aquatic adaptation has led to the nasal tubes from nostrils to throat becoming encased in bone, by development of a secondary roof to the mouth. The crocodile is now able while submerged to open its mouth without getting water down its windpipe. Hence in the crocodile the nostrils may be situated at the tip of the nose, while in those aquatic reptiles that do not have a secondary palate the nostrils tend to shift up to the top of the head.

The snout and jaws are fairly long in most crocodilians, especially in those living mainly on fish.

Intermediate forms between thecodonts and crocodilians are

**45** Comparison between skulls of a mesosuchian crocodilian, *Pelagosaurus* from the Lias (a–c), and a modern crocodile, the mugger *Crocodylus palustris* (d–f), seen from the side (a, d), from above (b, e), and from below, with lower jaw removed (c, f). The large skull openings in *Pelagosaurus* are primitive, and the internal nares (c, large opening in midline) are farther from the throat than in modern crocodile (f, pair of openings near hind end of skull). After Swinton.

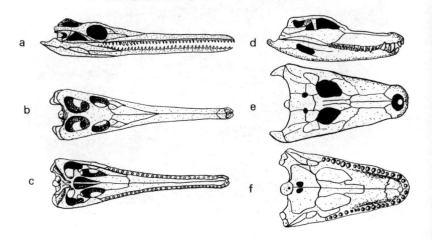

known from the upper Trias in the form of *Protosuchus* ('the first crocodile'). This small reptile still had a rather short head, but the arms and legs were quite like those of true crocodilians, and both the back and the belly were protected by rows of bony plates – a tendency to develop an armour of this kind is frequent in the Crocodilia. Protosuchus was contemporary with the 'false crocodiles' or phytosaurs, which have already been described, and which were now nearing the end of their career.

In the Jurassic the deployment of the crocodiles started in earnest. In the Lias we find true crocodilians, even though they still have some primitive characters in common with most other Mesozoic forms; it is especially noteworthy that the secondary palate was somewhat less developed than in modern crocodiles. These old-fashioned crocodiles, which are grouped in the suborder Mesosuchia, soon developed both long-snouted forms resembling modern gavials, and short-snouted, alligator-like types. Some of the latter were very small, only about a foot in length.

In contrast with other living reptiles the crocodile has a four-chambered heart; its blood circulation, therefore, is more efficient than in ordinary lizards, even if the condition is not quite as advanced as in birds and mammals.

Evidently the mesosuchians lived in much the same manner as the modern Crocodilia. They would inhabit lakes, rivers and the shelf seas, where they would lie awash with only the nostrils and eyes showing, looking like an old tree trunk, until some unlucky creature came too close. The more powerful forms would be able to overwhelm even very large land animals, if they could pull them into the water and drown them. The long-nosed forms, on the other hand, lived mainly on fish, and numerous such forms seem to have infested the shallow seas covering part of the European continent at the time, for instance during the Lias. A special group is formed by the true sea crocodiles or thalattosuchians of the Jurassic, which will be discussed in the next chapter.

Crocodiles of modern type (the suborder *Eusuchia*) appeared in the Cretaceous and reached their zenith in the Tertiary, when the

46 Skulls of the two largest terrestrial carnivores of all times, the dinosaur *Tyrannosaurus* (*left*) and the crocodile *Phobosuchus* (*right*), from the late Cretaceous in North America. Both animals were about fifty feet long; the skull of the long-snouted crocodilian measured six feet, that of the tyrannosaur four feet. *Phobosuchus* dates from the Belly River stage, while *Tyrannosaurus* comes from the Lance and thus was many million years later in time.

last mesosuchians gradually died out. The largest known crocodile was however a Cretaceous representative of the eusuchians, *Phobusuchus* or the 'horror crocodile', whose immense head was more than six feet long; the total length of this beast must have been some forty-five to fifty feet. In the flesh it probably resembled a large modern crocodile. It probably preyed on various kinds of dinosaurs.

## Flying lizards

The archosaurs were not too successful in their attempts to invade the marine environment, for not even the most high specialised marine ruling reptiles (the sea crocodiles) were able to colonize this environment permanently; thus the excursions of the archosaurs into the water tended to be restricted to an amphibious existence in shallow water (phytosaurs, sauropods, hadrosaurs, crocodiles). In the air, on the other hand, they were highly successful: they gave rise both to the birds and to the flying lizards. The latter form the order Pterosauria, which is the last of the five archosaurian orders.

The flying lizards appeared at the beginning of the Jurassic and lived almost to the end of the Cretaceous, but may have died out a little earlier than the last dinosaurs. The pterosaurs represent the first 'attempt' by the vertebrates to produce a flying animal, and if they look a little crude their existence for more than a hundred million years nevertheless is pragmatic evidence of adaptive success.

All the earliest pterosaurs belong to the suborder Rhamphorhynchoidea, the members of which had a long tail, in some cases with a small rudder at the tip. The wings of the flying lizards were formed by the arms, as in birds and bats; but while the flying membrane of the bat wing is supported by several fingers, somewhat in the manner of an umbrella, that of the pterosaurs was carried only by the immensely lengthened fourth finger (corresponding to our ring finger). The wings thus tend to be very long and narrow, as in a glider. In comparison with the finger, the upper and lower arm were quite short.

The three inner fingers, corresponding to our thumb, index and middle fingers, were free and protruded in front of the leading edge of the wing; they were clawed and may have been used by the pterosaur to hook itself up with when resting, and perhaps also in crawling or shuffling along the ground. The fifth finger was absent. A short bone extended from the wrist towards the body, supporting the part of the membrane that lay in front of the arm bones. The membrane also extended to the hind legs, but the feet were free of it; they carried five long toes with curved claws. Probably the pterosaur used them to cling to its perching place, perhaps also to pull or push itself over the ground; it was evidently unable to walk bipedally in the manner of birds.

The body of the flying lizards looks very small in comparison with the wings. The breastbone is fairly well developed and has a crest for the attachment of the flying muscles, but these elements are very small compared with those in birds, and it is clear that the pterosaurs were not as strong fliers as the birds. On the other

47 Flying reptiles. (a) *Nyctosaurus*, a crestless pteranodont from the late Cretaceous, wing span seven feet; (b) *Rhamphorhynchus*, a long-tailed pterosaur from the late Jurassic, wing span three feet; (c) *Dsungaripterus*, a short-tailed giant form from the early Cretaceous, wing span ten feet. Most flying reptiles have straight jaws but in *Dsungaripterus* they are curved upward, avocet-like.

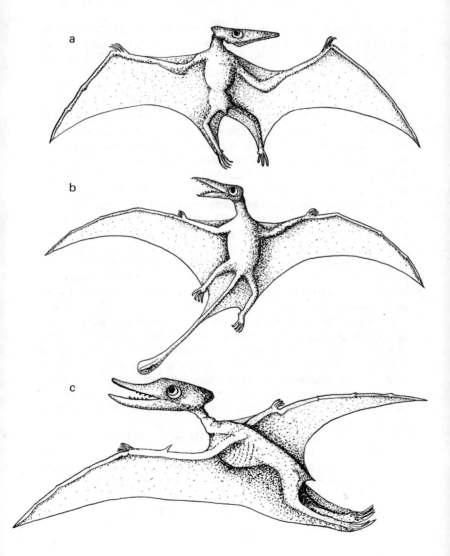

hand they may well have been proficient in the art of soaring, using the rising currents and the variations in the strength of the wind at different altitudes. As in the birds, many bones were hollow and probably contained air sacs, reducing the weight.

The head is very big in relation to the body. The jaws were well developed and carried numerous sharp teeth in the rhamphorhynchoids. In *Rhamphorhynchus* ('prow-beak') the jaws were long and slim, while the large, widely spaced teeth were pointing obliquely forward; this probably made it easier for the pterosaur to catch its prey on the wing. In *Dimorphodon* the jaws were deep and compressed; the long reptilian tail did not carry a stabiliser, whereas *Rhamphorhynchus* had a small, rhomboid rudder.

The brain of the pterosaurs was comparatively large and well developed, probably necessary to take care of the various difficult problems of coordination and balance connected with flying. The visual centres were very large, while the olfactory lobes were small, indicating that the sense of smell was unimportant in comparison with vision – another parallel with the birds. The similarity in bird and pterosaur adaptation is not surprising, since both have an archosaur ancestry.

Flying necessitates such intense and continuous activity and attentiveness that it is difficult to believe that an ordinary cold-blooded reptile could master it completely. It may perhaps be assumed that some degree of regulation of body temperature existed in the pterosaurs. Both the blood circulation, the breathing and the nervous systems were probably more highly developed than in ordinary reptiles, even if they did not attain the level of complexity seen in the birds.

The origin of the flying lizards is not exactly documented. *Dimorphodon* from the Lias is one of the first pterosaurs and has some primitive characters, for instance the long naked tail and the large and strong hands and feet, but it is already a typical pterosaur. Probably the flying lizards evolved from arboreal, jumping thecodonts, in which skin folds were evolved to help in gliding from branch to branch.

In the late Jurassic a more advanced suborder appeared, the Pterodactyloidea, in which the tail was almost completely reduced. The genus *Pterodactylus* comprises small forms varying in size between sparrows and thrushes. Only small, thin teeth remained in the front of the pointed jaws. Related forms with long jaws carrying closely spaced, bristle-like teeth are also known; this arrangement is not unlike the straining mechanism formed by the bills in some ducks.

Cretaceous pterosaurs tended to increase in size. A remarkable form from the early Cretaceous of China is *Dsungaripterus*, a large dragon-like creature with a wing span of about ten feet; there were teeth in the hind part of the jaws, but the front end formed a toothless beak that was curved upward as in the avocet and some other shore-living birds.

Still larger size was reached by the members of the pteranodont family, which culminated with enormous flying monsters like the Upper Cretaceous air dragon *Pteranodon* ('winged, toothless') which had a wing span of up to twenty-five feet. In the pteranodonts the big, pelican-like jaws were completely toothless, and evidently covered by a horny beak. The head, which looks enormous in relation to the small body, was made still larger in some forms by the development of a long crest, seemingly a counterweight to the bill, and with perhaps a steering function. In other pteranodonts the crest was lacking.

The history of the flying lizards comes to an end with this form; in the uppermost Cretaceous strata, where many kinds of dinosaurs are still found, the pterosaurs are missing.

How the pterosaurs moved on the ground is uncertain; at any rate it seems hard to believe that they were able to start from a flat surface. Probably they perched in trees, cliff niches and the like, where they could start by launching themselves out into the air. That many of the flying lizards, *Pteranodon* for instance, were marine forms and lived on fish appears probable, but one may well ask how they got into the air again after having caught a fish in the water.

48 Restoration of skeleton and wing membrane of *Pteranodon*, the largest-known flying reptile, with a wing span up to twenty-five feet. Its remains are found in the deposits of the Niobrara Sea, the great interior sea of the Late Cretaceous in North America. Its mode of life may have been somewhat like that of the present-day albatross.

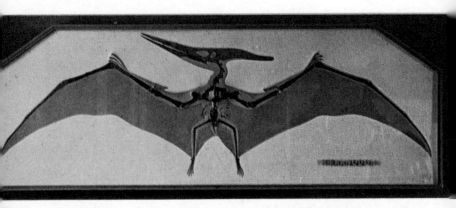

In being unable to walk bipedally the pterosaurs were significantly inferior to the birds. The same is true of the construction of the wings: a single tear of the membrane would result in an immediate crash, whereas a bird may lose several wing quills and still be able to fly. Perhaps it was this vulnerability, or perhaps some other factor which contributed to the extinction of the flying lizards in competition with the birds.

## Birds

We have now surveyed the entire archosaur group of the Jurassic and Cretaceous. Before looking at the other four-footed creatures of those times, we may take a look at those specialised offshoots of the archosaurian group, the birds.

The birds are generally not included in the reptilian class, but regarded as a class of their own, the Aves. Their most distinctive character is perhaps the possession of feathers, which certainly seem entirely different from the reptilian scales. Yet the feathers are in fact nothing but modified scales, and the annual moulting of the feathers is in principle the same thing as the reptilian moulting of the skin.

Otherwise, birds differ from reptiles mainly in those characters directly associated with flight. The temperature regulation (warm-bloodedness), the intense uptake of oxygen and nutriment and the efficient central nervous system are part of this adaptation just like the transformation of the arm into a wing.

The birds probably descended from Triassic thecodonts, which were already bird-like in many respects, for instance in their tendency to a bipedal gait and their hollow bone structure. Exactly how the birds first evolved is unknown, but the intermediate form between reptile and bird that is called *Archaeopteryx* gives clear evidence of the reptilian ancestry of the birds.

As early as 1860 an impression of a feather was discovered in the quarries at Solnhofen, a sensational find since all the birds then known were Cenozoic, while Solnhofen is Upper Jurassic. The following year a skeleton was found, excellently preserved with only the head missing; this was bought by the British Museum. In 1877 another skeleton was found, this time with the head intact; it is now in the Berlin Museum. A third skeleton was found in 1956. This specimen, now in the Erlangen University collection, is in somewhat poorer shape. In addition a feather, probably also belonging to *Archaeopteryx*, has been found at Sierra de Montsech in northern Spain.

*Archaeopteryx* was about the size of a modern crow, and the impressions of its plumage show that it was indeed a bird; but if only the skeleton had been found it might almost have been regarded as an archosaur. The tail, only a vestige in modern birds, was very long; the feathers lay in rows along its sides. The backbone was reptilian in structure; in modern birds the vertebrae are greatly modified, the back being very stiff and the neck extremely flexible.

The wings were still rather hand-like with long, free digits; in modern birds the fingers tend to coalesce. The head was reptilian with toothed jaws. In all modern birds the teeth are lost, and instead there is a horny bill, just as in the pteranodonts.

The *Archaeopteryx* was probably arboreal in habit, and the

49 *Previous page* Feathered skeleton of *Archaeopteryx*, the earliest bird, from the late Jurassic lithographic stone of Solnhofen, Bavaria. Two other skeletons (in London and Erlangen) of this reptile-bird are known, but the Berlin specimen shown here is the only one with the head intact. The skull has many reptilian characteristics, including teeth in the jaws, but the brain is larger than in reptiles of this size, an adaptation towards the motor coordination needed in flying. Note the feather impressions; presence of clawed fingers in wings; perching hind legs; and long,

wings may have evolved to help it move from tree to tree without having to descend to the ground, where many dangers lurked in the shape of dinosaurs and other reptiles. Probably the insulating capacity of the feathery plumage also assumed importance at an early stage in connexion with the evolution of an efficient regulation of the body temperature.

Occasionally a dead *Archaeopteryx* must have dropped into the river, which carried it out to the shore, where it finally in a semi-putrefied condition settled down in the lagoon where the future lithographic limestone of Solnhofen was being deposited. It is thanks to these unusual occurrences that we know anything about *Archaeopteryx*.

Sparse fossils from the Cretaceous period give us glimpses of the continued evolution of the birds. Many of the Cretaceous birds retained such reptilian characters as the toothed jaws, even though highly specialised forms·had already evolved. A remarkable line of evolution in the toothed birds can be traced from the early Cretaceous in Europe up to the late Cretaceous North American *Hesperornis* ('bird of the west') and led to the appearance of wingless, diving forms, in which the powerful, paddle-like hind legs were used for swimming. Other toothed birds show some resemblance to gulls and terns and probably led the same kind of life. All of these birds are found in marine deposits, and our knowledge of land birds is incomplete.

Most of the modern orders of birds seem to have been in existence towards the end of the Cretaceous. Unfortunately the material available for study of these primitive representatives of the modern birds is very fragmentary. Among them may be mentioned early members of the loon order (Gaviiformes) as well as heron- or stork-like birds (Ciconiiformes), the latter appearing perhaps as early as the beginning of the Cretaceous. There are also some types of wading birds belonging to the order Charadriiformes. The presence of many other modern orders in the Upper Cretaceous is suspected, but the fossil evidence is hard to come by. The lightly-built bird bones are easily shattered into unidentifiable reptilian tail. When, after some haggling about the price (finally set at £700), the first skeleton arrived in London in 1862, the Director of the Natural History Museum, Richard Owen, must have been somewhat shaken in his anti-evolutionary convictions; and the fact that the reptile-bird was a perfect example of a 'missing link' was confirmed by Professor Wilhelm Dames a quarter-century later, when the Berlin skeleton came into his hands. All the reptile-birds are now referred to the single species *Archaeopteryx lithographica*.

fragments. However, modern methods of collecting may lead to a great increase in our knowledge of the early history of the birds.

## Mammals

As we speak of the Age of the Dinosaurs, it may be worth while to recall that the mammals arose at about the same time as the dinosaurs and that thus there were mammals in existence throughout this Age. As early as the beginning of the Jurassic there were animals that may almost certainly be classified as mammals, and many of their predecessors from the Trias are border-line cases, sometimes regarded as mammals and sometimes as reptiles.

Modern mammals differ from reptiles in being warm-blooded; in having a four-chambered heart and a complete separation of oxygenated and deoxygenated blood; in having fur, or hair; and in suckling their young. None of these characters can be studied in fossils (except possibly the presence of hair, in special circumstances), so that it is necessary to use certain skeletal characters to decide whether we are dealing with a mammal or a reptile.

It has gradually turned out, however, that those skeletal characters distinguishing modern mammals from reptiles did not evolve all at the same time among the various transitional forms in the Mesozoic. The only solution then seems to be to select one key character. The structure of the jaw joint is now used by most investigators as the deciding factor.

In reptiles, the lower jaw consists of several bones, and the joint between the jaw and the skull is formed by two bones called the articular (in the lower jaw) and the quadrate (in the skull). The mammalian lower jaw, on the other hand, consists of a single bone, the big tooth-bearing dentary, and the jaw joint consists of a condyle on this bone and a socket in a skull bone, the squamosal. The articular and quadrate are none the less retained, but they have shifted into the inner ear to form auditory ossicles under the names malleus and incus.

This surprising transformation is illuminated by the fossil

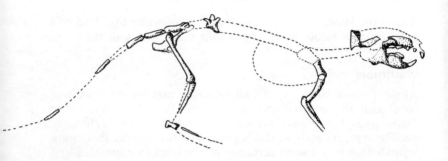

50 Fossils of Mesozoic mammals are so rare and fragmentary that for long only the teeth and jaws could be studied. Recent collecting has brought new evidence on the postcranial skeleton. This is the first skeletal restoration of a Mesozoic mammal, the multituberculate *Mesodma*, from the late Cretaceous of Montana. After Sloan and Van Valen.

record. In the transitional forms between reptiles and mammals both the joints were functioning at the same time – or, more precisely, they formed a single joint in which all the four bones took part. Gradually, however, the function tended to pass to the squamosal-dentary joint, making the old 'reptile bones' superfluous. Even when functioning in the joint they had probably played some role in sound transmission, since the jaw joint was very close to the ear in the mammal-like reptiles, and they were now free to function more efficiently in this role; so that it is easy to see that the change was highly adaptive.

The mammals are descended from the mammal-like reptiles (Therapsida), which flourished in the Permian and Trias. When the Trias was drawing to a close, only a few of these forms were still in existence, and they were all rather small. The best known are the tritylodonts, which survived in considerable numbers in the Lias. They had evolved a very specialised dentition. The front teeth or incisors had been reduced in number, so that only two upper and lower incisors were really functional, and these were enlarged as in modern rodents. The shape of the skull and the position of the cheek teeth also resemble these characteristics in the rodents, all of which gives a significant clue to the mode of life of *Tritylodon* and its relatives. This animal was about the size of a marmot. The tritylodonts became extinct in the Middle Jurassic.

At this time true mammals were already in existence in various parts of the world. However, all the Jurassic and Cretaceous mammals are small – the largest are no bigger than a cat – and the material is in most cases poorly preserved and incomplete.

The first Mesozoic mammals were discovered at Stonesfield near Oxford in the Middle Jurassic, and were described by Richard Owen. In 1854 Upper Jurassic mammals were found in the Purbeck fauna at Swanage on the south coast of England. O.C. Marsh discovered fossil mammals in the Upper Jurassic Morrison formation in Wyoming in the 1880s, and later on some teeth were found in the Wealden deposits from the basal Cretaceous in the Hastings area.

In 1939 W.G. Kühne initiated a new era in this research by a systematic study of Mesozoic fissure fillings. He selected areas known to have been dry land in the Jurassic, and with a limestone bedrock. In limestone there will be fissures and caves, which make an ideal environment for the fossilisation of small animals. Some of these are small carnivores which used the caves as dens, but the fossils may also belong to animals that fell into the fissures and were unable to get out. Kühne and his collaborators soon found a rich fauna of tritylodonts as well as true mammals.

Somewhat later the Americans introduced the method of washing and sieving large amounts of fossiliferous matrix. The results have been really splendid. Previously, only a few mammalian fossils of Upper Cretaceous date were known from the Lance formation and from Mongolia; now several productive sites are being worked, and a very large collection is awaiting study.

In the Jurassic there were various types of primitive mammals, among which the most important are called docodonts, triconodonts, and pantotherians. The docodonts, which are suspected to be ancestral to the primitive present-day Australian monotremes (the duck-billed platypus and the scaly anteater), were apparently small, insectivorous forms. The triconodonts were slightly larger, the largest being about the size of a cat. They were probably carnivores, preying on small reptiles and mammals. They survived to the early Cretaceous but then seem to have become extinct without issue.

The pantotheres are the most important Jurassic mammals, for they include the ancestors of all the living mammals except the above-mentioned monotremes. They were small or medium-sized, probably insectivorous in some cases, carnivorous in others. The history of the pantotheres may be followed up into the early Cretaceous, whereupon they are lost sight of. When we get back on their track in the Upper Cretaceous, the pantotheres have already given rise to the two big groups of mammals that are still in existence: the marsupials or pouched mammals (Metatheria) and the placental mammals (Eutheria).

The Upper Cretaceous marsupials are small, opossum-like animals. The placentals of this time are also small and may be referred to the order Insectivora in a broad sense. However, both carnivores and primitive ungulates are already foreshadowed among these creatures. Among the more truly insectivore-like forms we can distinguish between animals somewhat resembling hedgehogs (though we cannot tell whether they were spiny) and shrew-like or generally primitive types.

Beside these evolutionary lines another group of mammals must be mentioned, for it played an important role in the later Mesozoic and early Tertiary. These are the *Multituberculata*. They are the successors of the tritylodonts, as far as mode of life is concerned, although they probably did not descend from them. Predecessors of the multituberculates are known from the Jurassic, but this group became common only in Cretaceous times.

The multituberculates must be regarded as one of the most successful of all mammalian orders, if one reckons longevity a mark of success. It existed for more than 100 million years, which is more than any other order of Mammalia.

The multituberculates were rodent-like, just like the tritylodonts before them. They represent the herbivorous element in the Cretaceous mammalian fauna, and varied in size from that of a mouse to that of a terrier. The cheek teeth were highly specialised,

51 *Champsosaurus* was a large (six to eight feet) lizard, probably resembling a gavial or crocodile in the flesh, though in fact more closely related to the true lizards than to the ruling reptiles. It inhabited freshwater lakes and streams, and unlike dinosaurs survived into the early Tertiary. Based on a restoration by J.C. Germann.

forming in some multituberculates a highly efficient apparatus for opening husked fruits and seeds and chewing the pulp. We also know parts of their skeleton, and they turn out to be rather primitive and reptilian. The normal position was a squatting one with protruding knees and elbows. A recent restoration of *Mesodma* from the Upper Cretaceous in Montana shows a mouse-like creature some six or seven inches long.

This group survived with undiminished vigour into the earlier Tertiary and succumbed only when the competition from the true rodents became too great.

A modest size and a retiring mode of life were probably the best means of survival if you were a mammal and lived in the world of the dinosaurs. The big meat-eating dinosaurs probably were not even aware of the presence of mammals. On the other hand, the numerous smaller and medium-sized reptiles probably were a constant danger, which the mammals had to meet by developing greater speed and agility, a more efficient nervous system, a better method of reproduction and care of the young, and reliable control of internal temperature. It might be said that the mammals were tried and tested for 100 million years to make them ready to seize power at the end of the Mesozoic.

## Lizards and snakes

Present-day lizards and snakes are classed together with their extinct allies in the subclass Lepidosauria, which thus plays the main role in the modern reptile fauna, but was rather unimportant during the Mesozoic. Here belongs a very ancient group, the rhynchocephalians, which now survive in a small reptile called tuatara (*Sphenodon*) on some islands off the coast of New Zealand. Allied forms appear as early as the Triassic, and the Jurassic *Homoeosaurus*, which reached a length of about eight inches, was almost exactly similar to the tuatara.

A related group was the champsosaurs ('crocodile lizards') which appeared in the Cretaceous and survived in the early Tertiary. They were fairly large, amphibious lizards that probably resembled gavials, with a long narrow snout. They lived in fresh water and apparently fed mainly on fish. Except for the long jaws the head is broad and flat, with curiously closely-spaced eyes. The champsosaurs apparently managed well despite competition from the crocodiles, which lived in a somewhat similar manner, and like the latter survived the end of the Cretaceous, when so many other reptilians became extinct.

Modern lizards and snakes form a single order, Squamata ('scale-clad'). As early as in the Jurassic there are lizards resembling our lacertilians, and both the monitors and the iguanas are found in the Cretaceous. The earliest snakes appear in the lower Cretaceous, and later on in this period snakes somewhat like the living boas and pythons were common. No poisonous snakes are known from the Mesozoic.

The monitor group includes the marine mosasaurs, which will be treated in the next chapter.

## Turtles

It has been remarked by A.S. Romer *à propos* the turtles (Chelonia) that they are commonplace objects to us, only because they are still living; were they extinct, their shells would be a cause of wonder, being the most complete defensive armour ever found in a tetrapod.

Turtles basically like those of the present day were in existence as early as the Trias. The majority of the Jurassic and Cretaceous turtles belong to the suborder Amphichelydia, which dates back to the Trias. They had a shell similar to that in modern turtles, but the head and tail could not be drawn into the shell. Amphichelydians survived the end of the Cretaceous, and the last died out in the Pleistocene.

A more advanced suborder is that of the Pleurodira, which appeared in the later half of the Cretaceous. These are called side-neck turtles and hide their heads under the shell by simply turning them to the side. Only a few of the living turtles belong here, for instance the grotesque South American matamata.

The most advanced and successful turtles are the Cryptodira, in which the head is pulled back by bending the neck into an S. Most of the living turtles are cryptodires. The marsh turtles and terrapins represent a persisting little-changed central group that appeared in the later Jurassic and was quite common in the Cretaceous.

We thus have to visualise various kinds of turtles, but mostly amphibious shore and freshwater forms, in the Jurassic and Cretaceous environment. The marine turtles will be treated in the next chapter.

## Amphibians

Besides reptiles, birds and mammals the Mesozoic land fauna also comprised amphibians (Amphibia), as at the present day. Then as now the amphibians were rather small; we know little of them.

Among the salamanders (Urodela) we know allies of the modern ambystomids, now represented by some American forms, such as the axolotl. There are also relatives of the Proteidae (now including the American 'mudpuppy' and a few other forms) and Sirenidae.

Frog-like forms are known from the Upper Jurassic in Europe, Africa, and North America; they are usually regarded as an extinct family, the Montsechobatrachidae, related to a living family of frogs in which certain tail muscles are still present although the tail itself has been lost; this is a primitive character.

It may be assumed that the amphibians of the Mesozoic played the same modest role in nature as they do now, even if they showed a smaller range of adaptations – for instance, such highly specialised forms as the modern arboreal frogs probably were not yet present.

## Lung fishes

Among the many kinds of freshwater fishes that peopled the Mesozoic lakes and streams, the lungfishes or Dipnoi deserve special consideration. There were numerous lungfishes belonging to the family Ceratodontidae in various continents during the Mesozoic; this is the family to which the living Australian lungfish belongs. The typical genus *Ceratodus* has been found both in Europe, Asia, Africa, North and South America, and Australia. In the last-mentioned continent, however, the modern genus *Neoceratodus* takes over in the Late Cretaceous.

These fishes are fairly large, with a fringe like tail fin and pointed partly fleshy paired fins; the body is covered with large scales. These lungfish have a single lung. The modern Australian lungfish manages excellently in bad, oxygen-poor water that will kill all gill-breathing fishes, but cannot survive complete drying. *Ceratodus* is very close to the living genus and probably resembled it as regards mode of life.

# 5 Vertebrates of the sea

At all times it seems that the richly productive marine environment has attracted air-breathing vertebrates to return to the sea. The whales and dolphins, the seals, sea cows and sea otters among mammals, the penguins among birds, and the sea turtles and sea snakes among reptiles, are living groups which have, in the past, yielded to the lure. In Mesozoic times the reptiles dominated the fauna of the seas, as well as of the land. However, the marine reptiles generally belong to quite different groups: the almost complete absence of archosaurs is especially notable, except in part of the Jurassic period, when the sea crocodiles existed. Otherwise the Jurassic and Cretaceous seas were inhabited by ichthyosaurs, plesiosaurs, mosasaurs and sea turtles, as well as by fishes of many kinds.

## Fish-lizards

No other group of reptiles attained such an extreme degree of aquatic adaptation as the fish-lizards or ichthyosaurs (Ichthyosauria), which as the name indicates became very fish-like externally. They are the Mesozoic counterpart of the toothed whales and show about the same variation in size, from small creatures of common porpoise appearance up to cachalot-like giants more than thirty feet long.

The complete transition to life in the water necessitated a series of adaptive changes, and it is interesting to see how the ichthyosaurs solved these problems. In the first place the body would have to attain a compact streamlined shape without any externally visible neck, just as in the whales. For efficient swimming the backbone had to have a structure quite unlike that in a land animal, and this change too has been accomplished in the ichthyosaurs, whose vertebrae have a structure resembling that of the fishes. (An ichthyosaur vertebra is one of the things examining professors love to put into the hands of a budding palaeontologist.) With this type of backbone the fish-lizard was able to swim with a fast, 'long-wave' undulation of the body, instead of the slower 'short-

52 *Below* The structure of an ichthyosaur. (a) Skeleton and body outline as revealed by carbonised soft parts in certain well-preserved specimens, showing fusiform body shape, shark-like dorsal fin, and vertical tail fluke. The evolution of the tail fin may be traced from the primitive *Mixosaurus* (b) of the Trias, to the advanced type seen in (c) *Ichthyosaurus* of the Jurassic. This type of tail fin, in which the lower lobe is stiffened by the backbone, is termed hypocercal, and is common in sea-going reptiles of the Mesozoic era.

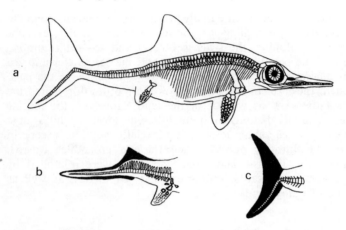

wave' undulation seen in eels and swimming snakes.

Efficient swimming also necessitates the development of a tail fin. Thanks to finds of ichthyosaurs in which the body contour has been preserved as a carbonised film, we now know that they possessed both a tail fin and a dorsal fin. The tail fin was in the vertical plane as in the fishes, but in contrast with the whales, in which the flukes are horizontal. In the ichthyosaurian tail fin it was the lower lobe that was stiffened by the backbone, which was sharply bent downward at this point, while the upper lobe lacked skeletal support. When ichthyosaur skeletons were first found with this peculiar crick in the tail, they were thought to be damaged, so that the creature was restored with a straight tail. How the tail fin evolved is shown by a comparison with the earliest (Triassic) fish-lizards, in which the backbone was only moderately bent and the tail fin consisted mainly of the upper lobe.

A tail fin like that of the ichthyosaurs is termed hypocercal; the opposite type of fin, in which the backbone turns up into the upper lobe, is termed heterocercal, and is found for instance among sharks. In swimming the hypocercal tail fin tends to force the rear end downward, thus causing the snout to turn upward; this effect

53 Skeleton of female *Ichthyosaurus* from the early Jurassic of Holzmaden, Bavaria, showing (*above*) young specimen inside rib cage and above stomach contents in body cavity; (*below*) enlarged detail of same. Length of adult specimen about seven feet.

can be counteracted by using the pectoral fins, whereby the total effect becomes a forcing down of the entire body into the water. It may probably be assumed that in the ichthyosaurs this was counteracted by the buoyancy of the body. The opposite arrangement may be seen in the sharks, which keep their heavy body afloat by swimming, using the combined lifting power of the tail fin and the pectoral fins.

The dorsal fin of the fish-lizards was high and pointed as in a shark or dolphin. In contrast with the whales the ichthyosaurs retained the hind legs, but they had been transformed, like the hands, into flippers or paddles. The phalanges of the digits, the wrist and ankle bones, and even the long bones of the arm and leg, had been transformed into more or less rounded, disc-like elements. In this respect, too, the Triassic ichthyosaurs represent a more primitive stage.

The head of the ichthyosaurs was big, with long, pointed jaws, armed with sharp teeth in most forms (although a few ichthyosaurs were toothless). They probably hunted fishes and other swimming animals. The nostrils were on top of the head just in front of the eyes. The eyes, which were very big, had a sclerotic ring consisting of small bony plates, perhaps as a support against the great changes in pressure during deep dives. Vision was important to the animal in hunting.

The structure of the fish-lizards clearly shows that they were completely dependent on water, and unable to emerge on land. It seemed natural to assume that they brought forth living young, and this was demonstrated by finds of ichthyosaurs with unborn embryos in the body cavity. In some cases it must be admitted that it looks as if the ichthyosaur had swallowed the young, but in other instances it is evident that these really are embryos, and there are even cases showing the young partially emergent from the body of the mother. The mother then died during birth, or else the labour was commenced immediately after the death of the mother – a phenomenon that may also occur in mammals.

The earliest ichthyosaurs appeared in the Trias; their ancestors have not yet been certainly identified, but a group of small aquatic reptiles from the late Palaeozoic (the Mesosauria) have been mentioned as possible candidates. The Jurassic was the apogee of the fish-lizards. In the early Cretaceous they were already less abundant, and in the Upper Cretaceous only a few types were left; they apparently died out a little before the end of this period, since they are absent in the youngest marine faunas of the Cretaceous. This seems odd, since both the mosasaurs and the plesiosaurs flourished up to the end of the Cretaceous, and the ichthyosaurs would appear to have been better adapted to life in the sea.

## Swan-lizards

The plesiosaurs, which appeared in the Late Trias, have also been called swan-lizards (Plesiosauria) because of the long, snaky neck seen in some species. Otherwise their appearance had very little of the swan about it, and Dean Buckland in his time compared the plesiosaur to a turtle with a snake strung through its body. The turtle part would then be represented by the short, flat body and the paddle-like limbs, while the snake would correspond to the neck and tail. However, all the plesiosaurs were not long-necked. The lengths of the head and the neck were inverse to each other, so that the head was quite small in the long-necked forms, and very large in the short-necked. The arms and legs were transformed into paddles resembling those of the sea turtles, but usually longer and narrower.

The swan-lizards' adaptation to a marine environment was evidently quite different from that of the fish-lizards, and they retained more of the original reptilian structure ('plesiosaur', 'nearer to the reptile'). It is sometimes stated that they 'rowed' or 'paddled' themselves about, but this is of course not correct if by rowing or paddling is meant that the oar is lifted out of the water between strokes. In actual fact the plesiosaur must have 'flown' through the water in the same way as the swimming turtles and penguins, using its paddles like wings.

54 Skeleton of *Trinacromerum*, a late Cretaceous (Niobrara Sea) plesiosaur from western Kansas. Related to *Kronosaurus*, this is a smaller and less extreme type of short-necked, long-jawed plesiosaur.

As in most other aquatic reptiles the nostrils were situated on top of the head just in front of the eyes. The jaws were armed with pointed teeth. In one group of plesiosaurs, the elasmosaurs, the neck became enormously long with up to 76 vertebrae. Probably it was usually coiled back as an S; when the prey came into range, the little head could dart out to catch it. The diet of the elasmosaurs probably consisted of smaller fishes and other swimming animals such as squids and belemnites, but they were

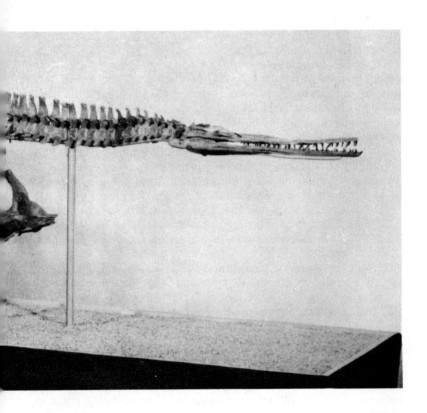

probably also quite able to catch flying animals that were unwary enough to come within their range, pterosaurs and sea birds. In some fossils the stomach contents have been partly preserved, and there are remains of pterosaurs, fish and cuttlefish. The animal also swallowed pebbles to help in breaking up the food in the stomach; in some cases up to several hundred such 'gastroliths' or stomach stones have been found in the belly region of a fossil plesiosaur. They have a dull surface, etched by the stomach juices (in contrast with popular opinion, which has it that the stomach stones should be shiny and polished).

The more 'normal' plesiosaurs, the family Plesiosauridae, were quite common in the Jurassic; in these the neck was moderately long. In the Liassic *Plesiosaurus*, which could be from six to fifteen feet long, the tail carried a small rhomboid-shaped fluke. Numerous well preserved skeletons of this form have been found at Holzmaden and in England. The last members of this family lived in the early Cretaceous; then dominance passed to the elasmosaurid family, which was well represented in the Upper Cretaceous seas. The largest forms, including *Elasmosaurus* itself, attained a length of 43 feet or more, more than half of which is head and neck.

A few predecessors of the Upper Cretaceous elasmosaurs have been found in deposits from the Lias and Lower Cretaceous. The 7½-foot *Brancasaurus* was such a miniature form, with only 37 neck vertebrae.

In another group of plesiosaurs, the pliosaurs and their allies, the head tended to become very big and long, while the neck was short with only a few vertebrae. The Liassic *Thaumatosaurus* still had a comparatively small head; this form measured some 11–12 feet in length. The evolution of the pliosaurs culminated in a tremendous monster from the Lower Cretaceous, *Kronosaurus*, in which the skull was three yards in length and is thus the largest reptile skull known. This animal was not unlike a sperm whale, and its mode of life may well have been analogous. Pliosaurs were still common near the end of the Cretaceous, but probably fewer in numbers than their long-necked cousins the elasmosaurs.

The first plesiosaurs appeared in the Upper Trias; their predecessors are not certainly known. A Triassic group of amphibious reptiles, the nothosaurs, is evidently allied to the plesiosaurs, and it is possible that the latter arose from some kind of nothosaur. The nothosaurs were small, rather long-necked and long-tailed reptiles; the limbs were only moderately modified for aquatic life, but some fossil finds have shown that the digits were webbed.

55 Skeleton of *Kronosaurus*, early Cretaceous short-necked plesiosaur from Australia; length about forty feet. The end of the tail was probably bent down to support a hypocercal fluke like that in early ichthyosaurs. It somewhat resembles a sperm whale, but unlike the whales retains four flippers; also, the tail fluke of a whale is horizontal, not vertical. Closely similar plesiosaurs have been found, for instance in the Late Jurassic of England.

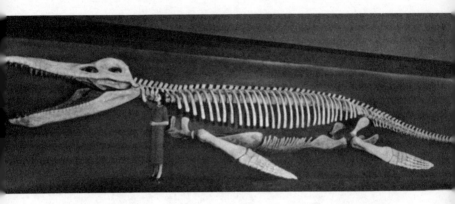

The nothosaurs could certainly move on land; whether the plesiosaurs were able to do this is unknown. Some of them probably could move about on land using their paddles in the same manner as a sea turtle or a seal, and they probably had to emerge to lay their eggs. It is striking that there are no finds of plesiosaurs carrying embryos, as in the case of the ichthyosaurs, and this supports the notion that they were oviparous and laid their eggs on land. Still, whether this was possible for such whale-like forms as the big pliosaurs seems doubtful.

### Sea crocodiles

As early as in the Lias we find crocodiles that probably were in part marine; these are mesosuchians belonging to the family Teleosauridae. Their long, slender jaws indicate fish-eating habits. In *Teleosaurus* the teeth pointed sideways out of the jaw, and those in the upper and lower jaws bit between each other. In *Mystriosaurus*, a big form up to forty-five or fifty feet in length, the teeth were vertical, but the front end of the snout was inflated. Remarkably preserved skeletons from Holzmaden show that the digits

56 Skeleton of *Tylosaurus*, a twenty-foot mosasaur or sea monitor from the late Cretaceous of Kansas. The great flourishing of the mosasaurs has been related to the proliferation of teleost fishes in the Cretaceous, probably their main source of food. Some mosasaurs, however, evidently specialised on shellfish, including ammonites.

were webbed and also reveal the stomach contents. The gastroliths seem still to bear traces of the 'ink' of cephalopods, thus indicating the staple diet of this crocodile.

The true thalattosuchians (Thalattosuchia), which lived in the Late Jurassic, may have descended from teleosaurs. In these sea crocodiles the mail had vanished completely, and the skull is more lightly built than normal in crocodiles. Both pairs of limbs are highly modified and resemble those of the seals; the legs are much longer than the arms and probably were important in swimming. In a specimen of *Geosaurus* from the lithographic shales at Eichstätt in Bavaria the contours of the body can be seen, revealing a tail fin resembling that of the earliest ichthyosaurs; as in these the vertebral column bends down into the nether lobe of the fin. However, the history of the sea crocodiles was not a long one, for only a few forms were left at the beginning of the Cretaceous, and soon after they also were gone. Their place was taken by other, possibly better adapted marine reptiles, perhaps by the mosasaurs, which became so numerous in the Upper Cretaceous.

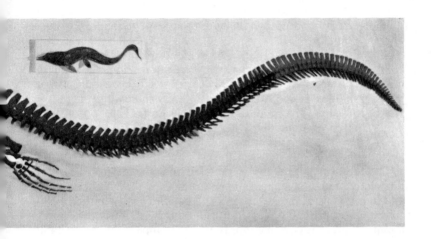

## Sea monitors

The mosasaurs or sea monitors (Mosasauridae) were great marine lizards, very closely related to present-day monitors. The numerous present-day species vary in length between ten inches and ten feet (as in the famous Komodo monitor). The mosasaurs grew still larger, up to 45–50 feet.

The history of the mosasaurs is restricted to the later part of the Cretaceous, but during this epoch they populated the seas in many guises. In the early Cretaceous some monitor-like lizards had become aquatic in habit, but they remained small; however, one of these groups evidently gave rise to the mosasaurs.

Like many land monitors the mosasaurs had a long, slim body and a laterally compressed tail. The arms and legs were much shortened, and the hands and feet formed flipper-like organs with webbed toes. In some mosasaurs, for instance *Clidastes*, there was apparently a tail fin like that in the thalattosuchians. The mosasaurs would swim with serpentine movements, probably coming nearer

57 Left side of ammonite (*Placenticeras*) showing marks made by upper teeth of a mosasaur, probably a member of the subfamily Platecarpinae. This was the seventh of a total of sixteen bites, ending with the mosasaur killing the ammonite and eating the soft parts. Late Cretaceous Pierre beds at Scenic, South Dakota.

to the popular concept of the great sea serpent than any other known animal.

The head was in general like that in monitor lizards but of course greatly enlarged. The jaws, which were more elongate than in monitors generally, were studded with teeth; and the lower jaw was hinged in the middle, thus greatly increasing the elasticity of the gape. The monitors are known to be able to bolt very large chunks of food, and the same evidently was true of the mosasaurs. The nostrils were situated on the upper side of the head, about half way between the eyes and the tip of the snout.

The mosasaurs give the impression of having been the most rapacious predators of the Upper Cretaceous seas. Various scars and mutilations of the lower jaws and paddles testify to their bellicosity. They probably hunted various types of freely swimming animals, and at least some mosasaurs were apparently in the habit of hunting ammonites – an extinct type of squid-like animal encased in a chambered shell, usually coiled in a spiral (see below). There is a highly dramatic find of the shell of an ammonite punctured by the repeated bites of a mosasaur, which on analysis (by E. G. Kauffman and R. V. Kesling) gives quite a detailed story of the encounter. It seems that the ammonite was attacked from above, and it is most likely that the mosasaur spotted it in the water below and dived down to attack it. It then bit across the shell several times, forcing it nearer its throat each time, but the shell was much too large to be swallowed whole, and after struggling for a while with this morsel the mosasaur let it go temporarily. When attacking again it crushed the living chamber of the ammonite in two bites, and probably pulled the animal out of its shell and swallowed it. The shell, now bearing the marks of sixteen successive bites, was left to sink to the sea floor and became fossilised.

The mosasaurs also include one group of forms (*Globidens*) with blunt teeth, probably feeders on shellfish and echinoids.

Did the mosasaurs emerge on land to lay their eggs? No juvenile specimens have ever been found, whereas adult skeletons of sea monitors are found everywhere in the world. It has been suggested (by

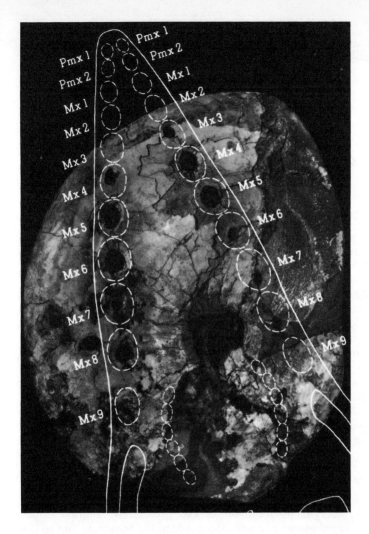

S. W. Williston) that the young lived in fresh water. He thought it possible that the female mosasaurs ascended the rivers to breed (perhaps giving birth to living young) and that the young remained in these more peaceful waters until able to fend for themselves. But, in the absence of fossil evidence, this must remain a field for speculation.

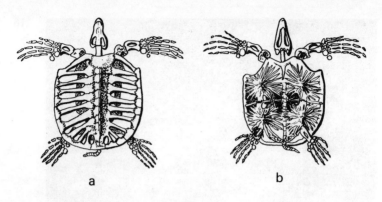

## Sea turtles

Like so many other reptiles, the turtles also returned to the sea. The biggest of the Mesozoic sea turtles are classified in the family Protostegidae, related to the modern sea turtles or Chelonidae. The protostegids are found in Upper Cretaceous marine deposits in America and Europe; they survived in the early Tertiary.

The shell was greatly reduced, and the hands and feet formed paddles used by the animal to 'fly' through the water. The protostegids include some very large forms: *Archelon* reached a length of twelve feet.

Members of the living family, the Chelonidae, are known from the Lower Cretaceous on. Examples of the modern genus *Caretta* (loggerhead turtles) appear in the Upper Cretaceous, while other contemporary forms are closely allied to the living *Chelonia*. The last-mentioned may also attain considerable size, with a shell length of up to five feet.

## Bony or true fishes

Among the bony fishes (Teleostomi) the ray-finned group, or Actinopterygii, predominate in the present-day marine fauna. They played more or less the same role in the Mesozoic, but to this group the Mesozoic era was a time of great transformation. Among the higher vertebrates, especially the reptiles and mammals, the major change occurred at the threshold of the Tertiary, when most of the reptiles became extinct and mammals became dominant. In the case of the actinopterygians an analogous revolution occurred many million years earlier, during the Cretaceous. As a

58 The great Cretaceous sea turtle *Archelon*, twelve feet long skeleton seen from above (a) and below (b). As in other sea turtles, the bony shell has been much lightened, and the limbs are adapted for swimming.

result, an essentially modern fish assemblage was in existence well before the end of the Cretaceous.

At the close of the Trias the fish fauna was dominated by primitive ray-finned fishes called chondrosteans or 'cartilage ganoids'. In spite of the name, these primitive ganoids had a well-ossified skeleton. The group was rapidly reduced in numbers in the course of the Jurassic and Cretaceous. The most primitive of these fishes, the Palaeoniscoidea, which date back to the Carboniferous and thus were very ancient by this time, finally expired in the Lower Cretaceous after some 200 million years of existence. At first sight they appear not to have differed too much from present-day actinopterygians, but the usually rhomboid scales were very thick (ganoid scales) and the tail fin had the same structure as in a shark, that is to say heterocercal with the backbone bent upward – contrary to the arrangement in the ichthyosaurs. The gill-cover also differed from that in modern fishes.

A somewhat more advanced group of chondrosteans, the Subholostei, appeared in the Trias and survived into the Lias; they are transitional to the holostean ganoids. The living sturgeons are relicts of the chondrosteans, and in these the ossification of the skeleton has been much reduced, whence the name 'cartilage ganoids' for the whole group. The first sturgeon-like fishes appeared in the Trias, and true sturgeons in the Upper Cretaceous, so that caviare has been available for about 100 million years.

Bone ganoids or Holostei are typical Mesozoic actinopterygians. They now survive only in the guise of a couple of relict forms, the garpikes (*Lepisosteus*) and the bowfin (*Amia*), both inhabiting freshwater basins in the New World. Garpikes appeared in the Upper Cretaceous, while the bowfins are somewhat older and date back to the Upper Jurassic. Among the numerous other kinds of holosteans that lived in the Jurassic and Cretaceous are found both long, slender, even swordfish-like forms as well as short, deep-bodied fishes resembling the sunfish, and of course a great variety of more normal-looking types. In general the holosteans differ from chondrosteans in having thinner scales, the

59 The fish within a fish. Excavation (*above*) and exhibition (*below*) of the skeleton of the large chirocentrid fish *Gillicus*, eaten by its still larger relative *Xiphactinus* (length sixteen feet and the largest-known fossil teleost). Other *Xiphactinus* specimens also contain remains of *Gillicus*, indicating that the giant fish habitually preyed on its smaller kin. Found in the Smoky Hill Chalk of the Niobrara beds, late Cretaceous, western Kansas.

tail fin is only supported by a short section of the backbone, and the cheek region is different in structure.

Some holosteans from the Jurassic and Cretaceous are still more advanced and are regarded as transitional to the teleosts, or modern ray-finned fishes. *Leptolepis* and its allies from the Trias, Jurassic and Cretaceous are often regarded as true teleosts. This little fish is especially common in the lagoon fauna of Solnhofen. Both in size and shape it resembles a small herring, and is evidently related to this group of fishes.

One of the oldest known present-day fish families is the chirocentrids, big tropical fishes resembling the herrings; they date back to the Jurassic. The living dorab (*Chirocentrus*) may reach a length of about four feet, but the Cretaceous *Xiphactinus* was much larger (482 centimetres in the largest specimen). A famous fossil specimen shows the complete skeleton of a related smaller fish (*Gillicus*), which had been gulped down by the *Xiphactinus* just before it died.

The true herrings (Clupeidae) appeared in the Lower Cretaceous. Some other modern fish families of the same age are the elopids, represented in the present-day fauna by a single, widely distributed species, and the chanids with a similar distribution. Otherwise the modern teleost families are often found to date from the Upper Cretaceous, and it seems evident that the main change in the fish fauna occurred between the Lower and Upper Cretaceous. The Upper Cretaceous seas were thus dominated by modern teleosts. Among the families appearing here may be mentioned the Albulidae, which includes some tropical fishes popular with sportsmen; the salmon (Salmonidae); the Osteoglossidae, tropical freshwater fishes; the Characidae, a group of very rapacious fishes including the notorious South American piranha; Mycrophidae, small deep-sea fishes; the Hemirhamphidae, odd-looking forms in which the lower jaw is much longer than the upper; the Holocentridae, a group of small, colourful coral reef fishes; the berycids, a family of spiny teleosts which was very common in the Cretaceous seas and is still important as a food fish; the thick-lipped Labridae; the Bramidae or sea bream; the Carangidae or pompanos and related forms. In addition there are many groups now extinct but closely allied to living families, for instance relations of the stomiatids (various remarkable looking deep-sea fishes), the carps, the sunfishes, and the eels. The Late Cretaceous fish fauna was indeed a highly varied one.

Besides the ray-finned fishes, lobe-finned fishes or Crossopterygii lived in the Mesozoic seas. They are fairly large fish with typical, partly fleshy fins; the Mesozoic forms are closely allied to the living *Latimeria*, which was discovered in 1938. Previously it had been thought that the crossopterygians became extinct at the close of the Mesozoic. A different group of crossopterygians gave rise to the land vertebrates at a much earlier stage.

## Shark-like fishes

The shark-like fishes (Elasmobranchii) played an important role in the Mesozoic seas. A typical Mesozoic group is formed by the hybodont sharks, which differ from modern sharks in some anatomical details, for instance in the structure of the jaws. The teeth tended to become differentiated into sharp teeth near the front of the mouth and blunt teeth further back; some of these sharks probably lived on shellfish.

In cartilaginous fishes it is unusual to find the whole skeleton fossilised, and this also holds for the hybodonts; most finds consist of teeth. In some hybodonts the teeth in the midline of the jaw did not fall out when they were worn, but were pushed back into the jaw as the new teeth emerged. The result is that we may find a long series of successive teeth, forming a characteristic looking whorl. It shows the gradual size increase correlated with growth: the innermost teeth in the whorl are very small, as they were formed in the early youth of the shark.

The hybodonts reached their apogee in the Jurassic; the last became extinct during the Tertiary. The present-day Port Jackson shark (*Heterodontis*) of the Pacific Ocean is a direct descendant of the hybodonts. A primitive characteristic is the position of the mouth at the point of the snout. Like the hybodonts it has two dorsal fins with stout spines. The genus dates back to the Jurassic.

A related group, the Notidanidae, also appeared in the Jurassic. These are fast, predaceous sharks, lacking fin spines and with a single dorsal fin; the tail fin is extremely large and lobate. Like *Heterodontis*, these forms are relicts of a much more numerous and varied Mesozoic group.

Beginning in the Jurassic and later on in the course of the Cretaceous, a succession of modern shark families enters the record. The mackerel sharks (Isuridae) appeared in the Jurassic; *Isurus* itself is known from the Lower Cretaceous, while the feared large white shark existed from the late Cretaceous. The spotted dogfish (*Scyllium*, family Scyllidae) dates back to the Jurassic; these sharks feed on small fishes and molluscs. The Carchariidae with the dangerous grey, blue, and tiger sharks also are of Jurassic origin. The same holds for the spiny dog-fish (Spinacidae) and the angel sharks (Rhinidae), flat skate-like forms probably transitional to the skates and rays.

In the Cretaceous are added the sand sharks (Odontaspidae) which include the goblin shark (*Scapanorhynchus*) originally found as a fossil, later discovered to survive in the Kuroshio current; the hammerheads (Sphyrnidae); the Scymnidae, and finally the saw sharks (Pristiophoridae). At the end of the Cretaceous, then, almost all of the main groups of living sharks were in existence.

The earliest skates of modern type also appeared in the Jurassic with the banjo fishes (Rhinobatidae) which are transitional between sharks and skates. They are still common in tropical and subtropical seas. In the Cretaceous we also find sawfishes (Pristidae), true skates (Rajidae), sting rays (Trygonidae) and eagle rays (Myliobatidae), in addition to extinct forms. The only important group not found in the late Mesozoic are the torpedo or electric rays (Torpedinidae), which appeared in the early Tertiary.

Finally, the family of the chimaeras should be mentioned. The Chimaeridae appeared in the Lias, but ancestral forms are known in the Palaeozoic. The chimaeras have a long rostrum; the mouth is ventral and armed with plate-like crushing teeth indicating a shellfish diet. The tail is shaped like a whiplash.

# 6 Flora and invertebrates

### The land flora

The flora was revolutionised during the Mesozoic, at an earlier time than the land fauna. When the Jurassic period opened, the flora was dominated by ancient forms, and flowering plants were completely absent; at the close of the Cretaceous, however, the flora was very like that of the present day.

The tree ferns had reached a climax in the later Palaeozoic, and were succeeded in the early Mesozoic by forest trees of more advanced types, but ferns and horsetails of large size were still common and in some areas covered wide expanses. Many of the fossil fern fronds from the Jurassic and Cretaceous in Europe are closely similar to those of exotic ferns now living in tropical and subtropical forests.

The Jurassic landscape, however, was dominated by trees belonging to the gymnosperms, represented today by the conifers. The Jurassic woods consisted, among others, of trees resembling modern cypresses, yew, and araucaria (monkey-puzzle). Especially common are forms related to the ginkgo or maidenhair tree, a well-known 'living fossil' often seen in parks. Another group of living fossils dating back to the Jurassic or even to the Trias are cycads. These are palm-like trees with an unbranched stem and cone-like flowers. They are dioecious, that is male and female flowers are on different plants.

Externally somewhat resembling the cycads, but with somewhat different organs of reproduction, and a branch-bearing stem, were the *Bennettitales* or cycadeoids, which are estimated to have represented about forty per cent of the forest vegetation in Europe. These trees, for instance the *Williamsonia* (see figure 61), thus formed an important element in the environment of the Jurassic dinosaurs. In *Williamsonia* the leaves were up to three feet long, and somewhat feather-like in structure with a strong central rib and a series of leaflets on either side. After their heyday in the Jurassic the Bennettitales dwindled in numbers during the Cretaceous, and became extinct well before the end of the period.

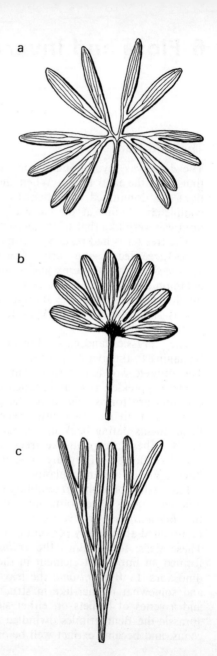

60 Leaves of ginkgo-like trees from the Jurassic of Siberia. (a), (b), two species of the genus *Ginkgo*; (c), *Baiera*. This group is today represented by the maidenhair tree, *Ginkgo biloba*, which has survived as a sacred park tree in China and Japan; its seeds are edible. The ginkgo group of gymnosperms is very ancient; trees of this type were in existence as early as in the Carbinoferous. After Heer.

The flowering plants or angiosperms appear at the beginning of the Cretaceous, but in general the Wealden woods were rather similar to the Jurassic, with cypress, araucaria, redwood (*Sequoia*) and *Ginkgo*, together with cycads and Bennettitales; the lower vegetation consists mainly of ferns. But we also find such new arrivals as oak, magnolia, plane-tree and laurel, with tropical forms like bread fruit tree and camphor tree. The origin of the angiosperms has been much discussed; as early as in the Jurassic pollen grains resembling those of angiosperms have been found, but the plants that produced them have not so far been identified. It has been suggested that the first angiosperms may have grown in dry regions, where little deposition and no fossilisation were taking place, and that they invaded the zones of sedimentation substantially later.

In the course of the Cretaceous period, the flora was gradually modernised, and the Upper Cretaceous vegetation is completely dominated by plants that are still with us. Here are found, for instance, walnut, poplar, hickory, birch, fig, willow, sycamore, maple, vine, ivy, sabal palm and many other modern forms, characteristic of the temperate and tropical woods. The last act in the dinosaur drama thus played in a completely modern setting.

## Terrestrial invertebrates

While we have a detailed knowledge of many of the invertebrates that inhabited the Mesozoic seas and coastal waters, little is known about those that lived on land. We know, for instance, that land snails probably existed, but they are little known. The many kinds of 'worm' groups that are found in the soil and in fresh water, have left practically no traces of their existence. About the arthropods we are better informed (insect faunas are known, for instance from the Lias in Switzerland and the Upper Jurassic in Solnhofen). There is no doubt that the Mesozoic woods and fields harboured immense swarms of arthropods. Scorpions and spiders, millipeds and centipedes existed well before the Jurassic, but those forms that

61 Mesozoic trees. *Below* (a) *Bjuvia*, a cycad from the end of the Trias in Sweden, after Florin. Cycads are still in existence. (b) *Williamsonia*, a member of the extinct Bennettitales, from the Jurassic, after Sahni. *Right* A present-day redwood grove in California; redwood forests were in existence well back in the Mesozoic.

lived in the Age of Dinosaurs are little known. Among the insects we find almost all the major present-day groups: from cockroaches through grasshoppers, termites, beetles, hemipterans (greenflies and bugs), lacewings, wasps and butterflies to the flies and gnats; but finds are comparatively rare and accidental. To try and judge the richness of the original fauna on this basis is something like attempting, on a cold winter day, to conjure up the myriads of crawling, jumping and flying creatures of the summer meadow from a few insect corpses in the window.

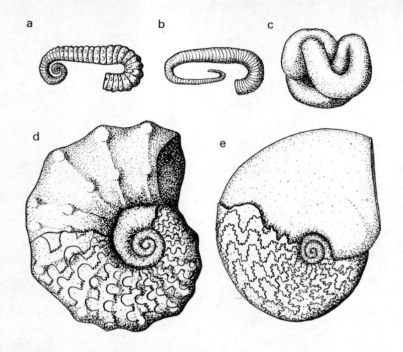

## Life of the sea: Ammonites

To us the dominant invertebrates of the Mesozoic seas were the ammonites (Mesozoic Ammonoidea), an extinct group related to the cuttlefish. Both groups are members of the class of animals called the Cephalopoda (a combination of words meaning 'head' and 'foot'). The name alludes to the fact that the molluscan so-called foot is in these animals united with the head and forms a ring of tentacles (often called arms) around the mouth. Another part of the foot forms a funnel, through which the animal can squirt out a jet of water (if necessary, mixed with 'ink') and thus propel itself rapidly backward to escape danger.

In the earliest cephalopods, the Palaeozoic orthoceratites, the animal was covered by a long, narrow, conical shell, the interior of which was divided into a series of chambers. Such shells are common in certain Palaeozoic limestones used as building stone in Scandinavia. At the open end of the cone there is a long chamber, in which the animal itself was concealed, with its head and tentacles emerging. The other chambers were filled with gas and with a fluid,

62 *Left* Aberrant (a, c) and planispirally coiled (d, e) types of Mesozoic ammonites. (a) *Ancyloceras*; (b) *Hamites*; (c) *Nipponites*; all mid-Cretaceous. (d) *Ceratites*, Trias; (e) *Oxynoticeras*, Lias. *Ceratites* shows early stage in lobation of suture line, *Oxynoticeras* a more advanced type. *Below* Structure of belemnite as restored (a) is dominated by the heavy skeleton; the chambered part contained a gas for buoyancy, while the compact hind part or 'rostrum' balanced the body of the animal. Most fossil finds of belemnites consist of the rostral part only, as in this specimen (b) of *Cylindroteuthis* from the Lias.

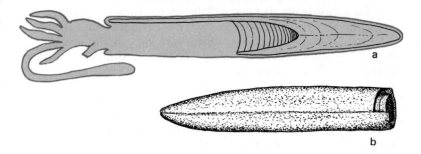

by varying the relative proportions of which the animal could regulate its weight in relation to that of the water.

As time went by, forms evolved in which the conical shell became coiled into a spiral, and this type still survives in the beautiful living animal called the pearly nautilus. The living *Nautilus* floats about with its shell coiled on top and the living chamber at the lower end. The genus was present as early as the Mesozoic.

The ammonites were in principle built like *Nautilus* and mostly had a planispiral shell too. Some details are different, however, one of the most important being that the walls between the successive chambers were not shaped as in *Nautilus*. In the last-mentioned form the suture between the chamber wall and the external shell forms a simple, undulating curve. In the earliest ammonites it was rather similar, but soon there appeared a tendency to lobation. As time went on, lobations of the second, third and even fourth order evolved. The result is that the suture lines in some ammonites may form an incredibly complicated, but at the same time often enchantingly beautiful pattern.

The outer side of the shell, which is smooth in the pearly nautilus, may often be richly ornamented in the ammonites, for instance with transverse ribs – branched or unbranched – or with knobs of varying size; some forms have a well-developed keel along the periphery. In some ammonites the resemblance to the spiral horn of a big ram is really striking, and the whole group has been named after the Egyptian deity Ammon, who was pictured with the head of a ram. On the other hand, there were also ammonites with a completely smooth shell. In many cases the mother-of-pearl layer is preserved on the shell, so that it can be admired in its original beauty.

During the Cretaceous many ammonites tended to become aberrant in shape, even though a main line with planispiral 'normal' ammonites persisted to the end of the period. In the aberrant forms appeared open spirals, U-shaped and even straight shells, and finally even completely irregular forms like *Nipponites*, which looks like a badly tied knot. Furthermore there were helicoid ammonites looking at first sight exactly like gastropods (*Turrilites*).

The variation of the form in the Cretaceous ammonites has fascinated many students, and the phenomenon has often been correlated with the subsequent extinction of the whole group. This is evidently true, although perhaps not exactly in the way that has sometimes been thought; the problem of the extinction will be discussed further in a later chapter.

## Belemnites and cuttlefish

The living cephalopods are divided into two subclasses: Tetrabranchiata with two gill pairs, and Dibranchiata with one pair. Both the pearly nautilus and the ammonites belong to the former division; but the other living cephalopods have only one gill pair, and are Dibranchiata.

In Jurassic and Cretaceous marine deposits remains of another group of dibranchiates, the Belemnoidea, are common. These are compact, cigar-shaped structures, very common in some chalk zones. The point of the 'cigar' was at the hind end of the animal, so that its name 'rostrum' is somewhat misleading. The fore part, which fitted into a conical 'alveolus' in the rostrum, was chambered and so corresponded to the chambered shell in the pearly nautilus and the ammonites. Finally, in front there was a thin dorsal plate. Some finds with preserved soft parts indicate that the animal possessed ten arms.

The belemnites were probably able to make quick reverse dashes by squirting water out of the funnel, like modern dibranchiates. Collisions were taken by the compact rostrum, which thus served as a protection for the chambered part. Some of the rostra show healed fractures that bear witness to collisions in the Mesozoic seas a hundred million years ago.

The belemnites flourished greatly in the Jurassic and Cretaceous; they survived in the early Tertiary, then died out. The living decapod (ten-armed) dibranchiates, however, may be descendants of early belemnites. In the sepioids, which include the common cuttlefish, the skeleton has become greatly lightened, in comparison with the belemnites, though the basic structure is similar. This group has been tentatively identified in Jurassic strata. Another group of cephalopods are the squids or Teuthoidea, which were quite numerous in the Jurassic and Cretaceous, and probably evolved in the Trias. They include many living deep-sea forms, notable among which are the giant squids.

In contrast, the Octopoda have no skeletal structures at all and hence are little known in the fossil state. An impression of an octopus in a late Cretaceous rock in Lebanon gives us our first glimpse of this group.

## Bivalves

The cephalopods treated above form a class within the phylum Mollusca. The phylum also comprises the classes Lamellibranchiata (or Pelecypoda) and Gastropoda, as well as some minor classes.

The lamellibranchs are molluscs with two valves (right and left) joined by a hinge at the back of the animal. Most live on the bottom in shallow salt or fresh water and move by crawling or burrowing, using the powerful, plough-like 'foot' (hence the name Pelecypoda, meaning plough-footed). A few are able to swim, for instance the

63 Mesozoic bivalves. (a) *Requiena*, a chamacean in which both the left and the right valve are coiled; early Cretaceous. (b) *Hippurites*, a rudist from the late Cretaceous. (c) *Inoceramus*, a common Cretaceous genus. (d) *Gryphaea*, a coiled oyster, Lias. (e) *Trigonia*, specimen from middle Jurassic; this is the only genus that is still in existence, but it is now very rare.

scallop (*Pecten*); this type was common well back in the Mesozoic.

Among present-day lamellibranchs may be found some fairly large, tropical forms classified as the Chamacea. They are a relict group, which was very common in the Jurassic. In the chamaceans the two valves may be very unequal in size, and one or both may become drawn out in a spiral. They inhabit shallow seas, coral reefs and the like.

Related to the chamaceans are the rudists, Rudistacea, a remarkable group of large, very highly specialised lamellibranchs, known from the Cretaceous only. In the rudists the right valve is conical, and in life was fixed to the sea floor; the left valve formed a small lid which closed the open end of the cone. The rudists only occurred in warm seas, and in Europe are found in certain zones of the Tethys; here they grew to form actual reefs. When a rudist died, a new one might settle into the hollow cone, and in some cases you may find a whole series of such conical valves threaded into each other. The valves are ornamented with strong logitudinal ribs, and the effect is such as to have suggested the name *Hippurites* (horsetail). The left valve, or lid, carried enormously large hinge teeth on the inside, forming a joint with the corresponding sockets in the right valve. An isolated lid of this type is one of the oddest things you can happen upon, and one that to the uninitiated looks utterly mystifying. The big rudists may be regarded as a kind of imitation of the big solitary corals.

They were frequently a foot high, and the largest known was six feet long and lay horizontally on the sea floor. Among other types of bivalves commonly found in the Jurassic and Cretaceous, the beautiful *Trigonia* species are noteworthy, with their characteristic trigonal shape and often strongly ornamented shell. They form important guide fossils in the marine Mesozoic. Originally desdescribed on the basis of fossils, the genus was later found still in existence in Australian waters, another case of a living fossil.

Oysters were common in the Jurassic and Cretaceous sea floors, both the modern genus *Ostrea* and extinct genera. Among the latter, the coiled oysters of the *Gryphaea* group are of special

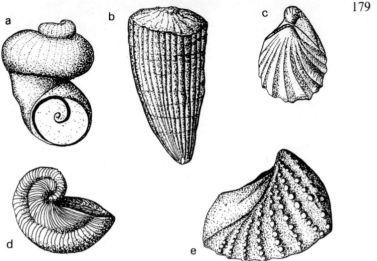

interest. The true oysters attach themselves to hard bottoms and hence have a flat shell. In *Gryphaea*, which lived on a mud floor, the lower (left) valve is coiled, so as to raise it above the soft surface of the bottom. Now several sequences are known in which you can see the *Ostrea* type gradually becoming transformed into *Gryphaea* with the passage of geological time – not only once, but at different times! This has long been regarded as a perfect example of iterative (repeated) evolution of one and the same type.

More recent studies seem to suggest that it was less a genetic change than a modification of the style of growth, resulting from the type of substratum. If the spat settled on a hard bottom, it grew up to be a flat, *Ostrea*-type oyster; the softer the bottom, the greater the coiling induced. This is dramatically illustrated by the young *Gryphaea*, which on a soft mud bottom happened to settle down on the only available hard area, the valve of a dead *Gryphaea*, and grew up to look like an *Ostrea*.

At a time when the mechanism of evolution was less understood than now, *Gryphaea* was sometimes taken as an example of compulsive evolution, in which change was forced in one direction regardless of its adaptive value, so-called orthogenesis. *Gryphaea* was a particularly sensational case, since the compulsive evolution would finally lead to a form in which the left valve was so highly coiled that its umbo would press upon the right valve and the animal become immured. At this stage the population would succumb, killed by orthogenesis. The reality is utterly different from this nightmarish conception, for even in the most coiled specimens the borders of the valves and the hinge become modified in such a way that the valves may open whatever the degree of coiling.

Among modern lamellibranchs, the freshwater *Unio* is well known in the Mesozoic; the same holds for the cockles (*Cardium*), the genera *Lucina*, *Tellina*, and many others. Razor clams (Solenidae) appeared in the late Cretaceous; they are one of the youngest lamellibranch families.

## Gastropods

Then as now, the gastropods (Gastropoda) formed an important part of the marine fauna, being especially common on the beaches and on the floor of shallow seas. The cap-shaped *Patella* (limpet) was a common sight in the tidal zone, on rocky beaches, even as it is now; and forms closely allied to the living winkle, *Littorina*, were also present. A peculiar extinct form, more distantly related to the winkles, is *Aptyxiella* of the Jurassic and Cretaceous, whose shell was an immensely steep and tall helicoidal spiral. In complete contrast to this are the almost planispiral, externally somewhat ammonite-like gastropods of the *Talantodiscus* type. The last-mentioned is closely allied to a present-day deep-sea form, *Pleurotomaria*, which was also common in the Mesozoic seas and is one of the most ancient and primitive of all gastropods.

Very common and of great importance as guide fossils are the fairly large, often colourful *Strombus* shells, armed with long spines; they are now found in tropical seas, and are often used for ornamental purposes. The beautiful cowry shells are of slightly more recent date, and appear only in the Cretaceous. In comparison with these, the purpura and its relatives (Muricidae) are rare in the Cretaceous, and their heyday began in the Tertiary.

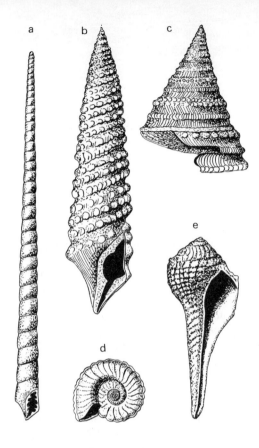

64 Mesozoic gastropods. (a) *Aptyxiella*, middle Jurassic, an exceptionally tall member of the extinct superfamily Nerineacea; (b) *Nerinea*, late Jurassic, same superfamily; (c) *Pleurotomaria*, late Jurassic and (d) *Talantodiscus*, Lias, members of the long-lived Pleurotomariacea ranging from Cambrian to Recent; (e) *Medionapus*, representing the progressive gastropods that arose in the Cretaceous (Volutacea and other superfamilies).

## Arthropods

The arthropods (Arthropoda) comprise the living crustaceans, spiders, insects, and their relatives. The terrestrial forms have been discussed above. The marine arthropod fauna of the Jurassic and Cretaceous is basically similar to that of the present day. Among crustaceans we already find crayfish, lobster, crabs, shrimps, and wood-lice; the crabs seem to have evolved at about this time, and enter the scene in the Jurassic. The ostracodes, small bivalved crustaceans, were common in the Mesozoic seas, as in those of our days, and are of great importance to stratigraphers. Barnacles resembling the living forms were also in existence; they grow attached to rocks and other hard objects on shallow sea floors. Most other main types of crustaceans probably existed in the Mesozoic, but we know little about their fossil history. In general the arthropods, with the exception of some forms like trilobites and ostracodes, are rarely well enough preserved to have a good geological record.

## Echinoderms

The 'spiny-skinned' animals or Echinodermata are a phylum that has played a most important role in the fauna of the seas ever since the early Palaeozoic. In our day it comprises the sea-urchins, starfish, brittle stars, sea cucumbers, and crinoids or stone-lilies and feather-stars. In addition there are various extinct classes, all of which however died out before the end of the Palaeozoic. Thus the Jurassic and Cretaceous echinoderms belong to living classes, and evidently resembled their living descendants in mode of life.

A common character in the echinoderms is a tendency to five-rayed symmetry, and an internal skeleton formed of calcite; most have a well-developed water vascular, or hydraulic system. The most primitive living group is the Crinoidea, comprising the stone-lilies and the feather-stars. Both were present in the Mesozoic.

The stone-lilies are attached to their substratum by means of a flexible stem, a column consisting of a great number of round or polygonal discs. The animal itself consists of a calyx or cup with the mouth turned upward and surrounded by five simple or branching arms. The stone-lilies were common in the Mesozoic, and some of the most exquisite fossils known represent this group, for instance in the Lias of Holzmaden. They look as if about to sway gracefully with the next movement of the water.

In the feather-stars there is no stem, and these animals move freely about, swimming in the water or crawling on the sea floor. They are now much more common than the stone-lilies, but in the Jurassic, when they first appeared, the relationship was the other way around. In the course of the Mesozoic, however, the free-swimming crinoids increased greatly in numbers. The Jurassic

65 Fossil sea lilies, *Pentacrinus subangularis*, from the early Jurassic of Holzmaden, Bavaria. Like almost all other crinoids after the Palaeozoic, *Pentacrinus* belongs to the subclass Articulata, in which the 'cup' joined to the stem is very small, and the main part of the calyx is made up by the lower ends of the arms, being connected by plates to form a flexible integument.

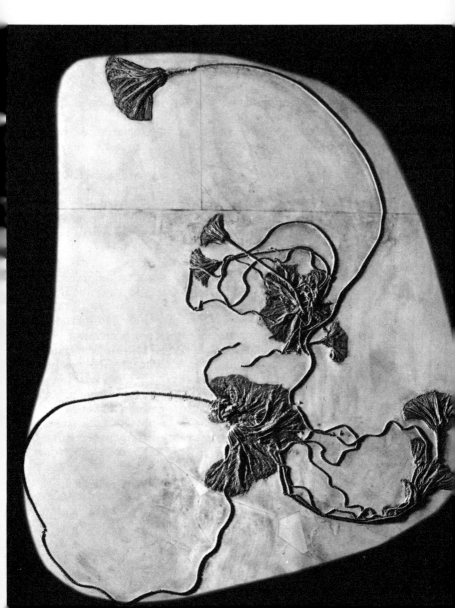

*Saccocoma* was a peculiar form with leaf-like, paired expansions on its arms, which may have made it an efficient swimmer.

The sea-cucumbers, Holothuroidea, generally leave only their small, sometimes microscopic skeletal elements as tokens of their existence; typical Mesozoic elements are cross-, wheel- and hook-shaped, while the beautiful anchor shape (*Synapta*) becomes common only in the Tertiary.

Starfishes (Asteroidea) and brittle stars (Ophiuroidea) were numerous in the Mesozoic seas, and have left beautiful fossils; the Mesozoic types are fairly close to the modern.

The sea-urchins, Echinoidea, which have a hard and resistant test ('shell'), have an excellent fossil record. In the Mesozoic, the 'regular' sea-urchins were especially numerous; in these forms the five-rayed symmetry is persistent. The body is often round as a ball. Urchins of the *Cidaris* group are very common; they carry very big, sometimes javelin-like spines. However, the spines may be lost after death and then one finds only the round tests; the strong ornamentation of the plates which make up the test may make them very beautiful. Sea-urchins of this type crawl about on the bottom with the help of their spines.

In the other main group of sea urchins, the 'irregular' ones, the five-rayed symmetry has been lost, and they are bilaterally symmetrical instead, with a right and left side. The irregular urchins usually live as burrowers in the bottom sediment. The first irregulars appeared in the Jurassic, and their ranks swelled during the later Mesozoic, until at the end of the Cretaceous the group comprised all the four orders still in existence.

One of the earliest good examples of species evolution was discovered towards the end of the nineteenth century, when Rowe investigated the irregular urchin genus *Micraster* so common in the Chalk. He was able to show that several different characters, for instance the shape of the test, the ornamentation, and the position of the mouth, underwent gradual change as you ascended in the series of deposits. The final result was a form so unlike the ancestral one that it had to be regarded as a different species.

66 *Right, Micraster,* an irregular sea urchin from the Cretaceous, lower and upper views. This genus, which comprises many important guide fossils in the Chalk, was early noted for its rapid evolution, and provided one of the first sequences of evolutionary change discovered by science. *Below,* Finds of marine fossils in present-day mountains, such as this sea urchin (*Miocidaris*) from the mid-Triassic Monte San Giorgio in Ticino, Switzerland, point to the paradoxical fact that mountains originate from the sea floor. This urchin belongs to the Regularia, and was armed with large spines.

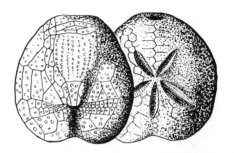

67 Mesozoic corals. The solitary forms (a) *Montlivaltia* and (b) *Oppelismilia* show the basic plan of the cup-shaped individual or corallite, with its radiating septa. The type is retained in colonial reef-builders like (d) *Isastraea*, in which corallites are still separated by walls. In (c) *Thamnasteria* the walls have disappeared so that the septa of adjacent individuals unite. All are Jurassic: (a) Lias; (b, d) Middle Jurassic; (c) Oxfordian, Upper Jurassic.

## Lamp shells

The Brachiopoda or 'arm-footed' is the somewhat curious name of a phylum of marine shelled animals, which at first sight look somewhat like mussels. The inner anatomy, however, is completely different from that of the molluscs, and the valves are not right and left valves but dorsal and ventral ones. They are opened and closed by powerful muscles. (Lamellibranch valves are closed by muscles, but opened by the tension in a ligament, so that a dead mussel will usually open up, in contrast with a dead brachiopod.) The hinge line between the two valves may be long or short. Behind the hinge there usually protrudes a stalk, with which the brachiopod attaches itself.

The actual animal fills only part of the space between the valves; a large part is taken up by the looped or spirally coiled 'arms', which carry the food particles to the mouth. Some brachiopods bury themselves in the bottom sediment, others live on rocky sea floors. The number of living species is estimated at about 260, so that the importance of the lamp-shell phylum in the modern fauna is very modest; but many thousand fossil species have been discovered. The brachiopods are thus highly important in palaeontology, as guide fossils and for the study of evolutionary problems.

The heyday of the brachiopods was in the Palaeozoic; in the Jurassic and Cretaceous most of the old groups are either extinct or greatly reduced in number. This is probably correlated with an increase in the number and diversity of competing molluscs, especially bivalves. Among the oldtimers that survived were, however, some types of brachiopods that survive even today and seem to be very like their Cambrian or Ordovician predecessors. They belong to the class Inarticulata, in which there is no true hinge between the valves. Both the burrowing *Lingula* and the small rounded *Crania*, which prefers a hard, rocky bottom, are 'living fossils' which have been identified as early as the Ordovician.

The predominant brachiopods of the Jurassic and Cretaceous seas were hinged forms (class Articulata) of the *Terebratula* and

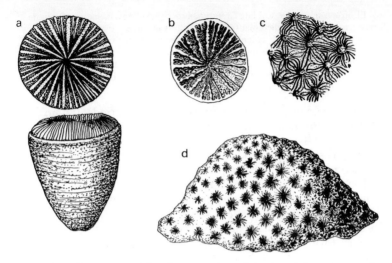

*Rhynchonella* groups. The first-mentioned is the richest in species among living brachiopods; these are forms with rounded, smooth or radially striped valves with a short hinge-line. On the inner side of the dorsal or brachial valve the terebratulids have a well developed skeleton, which is called a brachidium, for the support of the 'arms'. In the North Sea a species of *Terebratula* may be found living on stony bottoms.

The name *Rhynchonella* suggests a small bill, and alludes to the beak-like part of the lower shell that protrudes at the back, behind the extremely short hinge-line. The genus survives in the modern fauna, for instance in the North Atlantic. Both groups comprise important guide fossils in the Mesozoic.

## Corals

Both reef-forming and solitary corals (*Anthozoa*) were found in the Mesozoic seas. The ree corals have stringent environmental requirements: the temperature of the water must be high (not below 20°C), and the salinity must be normal; the depth should be no greater than 150 feet, and the water must be very clear. The coral animal lives in symbiosis with an alga, which lives in its body tissues and is dependent on sunlight. That corals lived much farther north in the Mesozoic than they do now indicates the warm climates of those times. By studying the geographic distribu-

68 Electron micrographs of late Cretaceous coccoliths, limy skeletons of minute green flagellates (one-celled plants). Serious study of these extremely abundant organisms has begun only recently; they promise to become very useful stratigraphic tools. There is no shortage of material, for some marine sediments contain up to 100,000 coccoliths per cubic millimetre. Specimens come from Dallas, Texas, Lake Waxahachie, Texas, and Kolby Gaard, Denmark; magnification varies from $x$ 10,000 to $x$ 30,000.

tion of fossil coral reefs dating from different periods, the position of the equator at successive times may be approximately estimated, and such 'palaeo-equators' will give clues as to the movements of the poles relative to the continents. They have been shown to be in good agreement with the palaeomagnetic data.

The primitive types of coral that had populated the Palaeozoic seas – the tabulates and rugose corals – had died out before the beginning of the Jurassic, and had been succeeded by a new series of coral forms, the Hexacoralla. Their diversity was considerable; there were solitary forms like *Montlivaltia*; simple, branching colonies like *Thecosmilia*; and tightly packed, reef-forming corals like *Thamnasteria*. The coral reefs grew only in a limited zone off the coast. Analysis of the structure of the reef, the position of its younger parts in relation to its older, etc., makes it possible to determine the oscillations of sea level and many other problems. Study of fossil coral reefs and the processes within them has become almost a science in itself.

## Other invertebrates and plankton

Little can be said here about the remaining invertebrates. In some instances they have a good fossil history but are only likely to interest the specialist – this is true for the Bryozoa (or Polyzoa), small colonial organisms forming rather inconspicuous incrustations on the sea floor. In other cases the fossil record is poor or entirely lacking, for instance as regards the numerous 'worm' phyla and most of the sponges.

Among the one-celled animals certain groups have a durable test and are easily fossilised. Most important among these are the foraminifers, which are of the greatest significance in stratigraphy. The oil geologists in particular rely greatly on the study of foraminifers, for they frequently have to study the sequence of rocks on the basis of bore cores, and the only fossils to be found in sufficient numbers in the restricted volume of such a core are micro-fossils such as these.

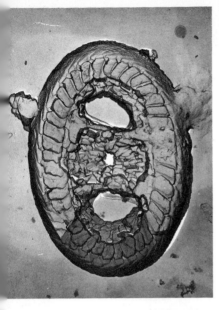

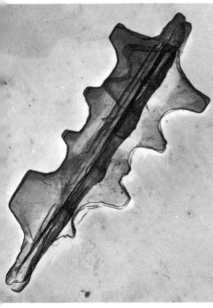

Some sedimentary rocks are almost exclusively composed of foraminifer tests. Certain types of foraminifers reach very large size in comparison with other one-celled animals; we have already discussed one such group of foraminifers, the late Palaeozoic fusulinids. They were succeeded by other types of 'giant forams' in the Mesozoic seas.

In addition to the foraminifers, among unicellular animals only the radiolarians have a well-documented fossil record; but they have not yet been studied in similar detail. Both groups still form an important part of the plankton of the sea – the assemblage of floating forms that is wafted about by the ocean currents. Planktonic plants of course also existed in the Mesozoic seas, but the only ones that have left a fossil record are the coccolithophorids, a group of flagellates with a limy skeleton; they form a major constituent in the Chalk.

# 7 Geography of the Jurassic and Cretaceous

### Europe in the Jurassic period

The geography of Europe at the beginning of the Jurassic represents a transgressive phase in the long, shifting combat between water and land characteristic of the history of this continent. On one hand there are the highlands that were formed in Palaeozoic times during the Caledonian and Variscan orogenies, and in earlier times; on the other, is the great Alpine geosyncline, the Mediterranean of this period, which has been named Tethys. From analogous geosynclines in the early Palaeozoic, the Caledonian and Variscan ranges had been born in a series of orogenic phases, the last of which occurred in the Permian. Here belong the Scandinavian and Scottish mountains; a second arc extends from southern Ireland through southwest England and the Ardennes to the Harz; and finally, a third chain is formed by the Armorican massif of Brittany, the French Central massif, the Vosges, Black Forest, and Erzgebirge. Together with still older land areas these highlands formed the continent that lay to the north of the Tethys. The lowlying plains of this continent were gradually inundated during the Mesozoic transgressions. The strata deposited in these shelf seas are mostly little disturbed, while those of the geosyncline show folding, overthrusting, and sometimes metamorphosis induced by high temperature and pressure.

South of the Variscan continent the development of the Tethyan deep got under way, perhaps in the Permian; in the Trias it was fully developed, and grew to ever mightier proportions in the Jurassic. Now the future arc of the Alps was fully visible in negative relief, as a trench. Eastward, this great Mediterranean gradually extended throughout Asia to the Pacific, marking those tracts that in future were to buckle up and form the Alpine ranges: the Alps, the Carpathians, the Caucasus, the Himalaya.

The Liassic geography of Europe is illustrated by the map. The Tethys may be seen in the south, extending over the Mediterranean countries of the present day; out of the deep, where folding has already commenced, an archipelago of mountainous islands is rising. South of the Tethys, Africa was emergent, but in the north the sea was flooding the lower parts of the continent, forming a shallow sea, connecting with the Boreal Sea (a predecessor of the Arctic Sea) by way of the North Sea basin. To the northeast, Scandis and the Russian platform were united to form a continent, including all of the Scandinavian peninsula except Scania at the southern tip of Sweden, which was inundated from time to time.

In the middle of the seaway extending to the north from Tethys there was a big island, the Brabant Island, comprising eastern England and Belgium. Further to the north, parts of northern England and Scotland were dry land, and formed the Scottish-Pennine Island. The Central massif in France formed a large, mountainous island.

The situation in the west is less clear, because much of the evidence is out of our reach beneath the sea. The palaeogeographers, however, postulate land here. But was this a moderate-sized land, extending from Spain to Ireland, 'Armorica', or a great North Atlantic continent, the North Atlantis? In any case it is evident that large areas in western Europe were dry land: the greater part of Ireland, Wales, Cornwall, Brittany, and the Spanish Meseta.

The inhabitants of the Liassic shelf sea are well known to us, thanks to the rich harvest of fossils, especially from the black shales of Holzmaden in southern Germany and the Liassic beds of the French and English Channel coast, particularly in the Caen area and Dorset. This sea is dominated by the ichthyosaurs, plesiosaurs, and crocodiles; besides, great swarms of the ammonites, belemnites and other invertebrates of the Liassic seas are found. We also find some fragments of land animals, for instance the stegosaur *Scelidosaurus*, and remains of flying lizards. In Tethys the fauna is somewhat different with typical 'alpine' ammonites (*Phylloceras*) and Mediterranean elements like the very common brachiopod, *Terebratulina aspasia*. While coral reefs are present in the shallow zones of Tethys, corals are rather rare in the shelf sea of the Lias. Perhaps this is partly because the depth was too great for corals, perhaps also because the water was too clouded by mud particles; the Lower Jurassic was not a friendly environment for corals. The open seaway to the Boreal Sea may also have effected a certain cooling of the Central European shelf sea.

During the millions of years that followed the Liassic epoch, the seas tended to recede. At one time in the Middle Jurassic, for instance, the Scandian land extended across Denmark westward almost to the present-day coast of England, and the corridor from the Boreal Sea to Tethys was now narrower and affected by masses of silt and mud carried to sea by great rivers from Scandis, from the Scottish-Pennine Island, and perhaps from North Atlantis between Ireland and Scotland. Some palaeogeographers assume that a Cimmerian land stretched all the way across the North Sea from Scotland to Scandinavia in the Middle and early Upper Jurassic.

Whether because the climate became warmer, because the cool waters of the Boreal Sea were now expelled from the Central European shelf, or because the sea became shallower – whatever the reason, corals now proliferated in many parts of Europe. Some fossiliferous deposits, for instance at Stonesfield near Oxford, give glimpses of the land animals that populated the Brabant Island and the other islands in the European archipelago; both sauropods, theropods and stegosaurs appear among these dinosaurs, but we also find fragments of the small mammals that scurried about in the shade of the fern thickets or hid in the foliage of the ginkgo trees and the bennettitales. In the sea, the regime of the ichthyosaurs, plesiosaurs, and sea crocodiles continues.

In the Upper Jurassic, most of the Russian Platform was flooded, while Scandis remained high and dry; its southwestern shore oscillated in Zealand and Jutland. In the Central European shelf sea, the regression continued, and coral reefs are now common. The Brabant Island was growing in size, extending ever further to the east, finally reaching into Poland; south of the Central massif in France a peninsula corresponding to the present-day

69 Mesozoic geography of Europe. *Left*, Lower Jurassic; *right*, Upper Cretaceous. Southern Europe, situated in the Tethyan geosyncline, shows continuous movements in the labour preceding the birth of the Alpine mountain ranges. North Atlantis and Scandis form stable blocks to the north-west and north-east. In the roughly triangular region between these areas there is an alternation between transgressions and regressions.

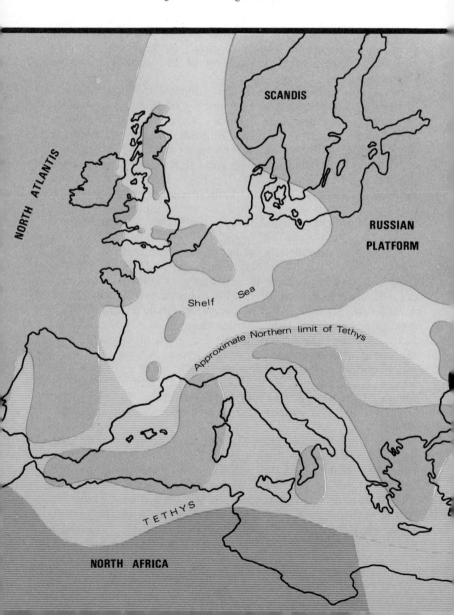

Pyrenees emerged, connected with the Armorican land in the Bay of Biscay. Another peninsula was formed by the persistent Spanish Meseta in the southwest. Brittany, Cornwall, Wales and Ireland remained as parts of Armorica or North Atlantis, and the coast line tended to advance to the east.

The most remarkable of the Upper Jurassic fossiliferous sites is Solnhofen in Bavaria. Here are found atolls, circular sponge and coral reefs enclosing lagoons that become silted up with sediments forming an exceptionally pure, fine-grained limestone. Generation after generation of quarrymen have been working this limestone, which is used for lithographic purposes, and a great many fossils have been discovered and collected.

In this fine, limy ooze, even impressions of such flimsy organisms as jellyfish have been preserved, to say nothing of the soft parts of higher animals. Here the skeletons of the *Archaeopteryx* were preserved together with the impressions of their plumage, and there are also fossil pterosaurs where the thin wing membrane can still be seen. Besides there are many crustaceans, ammonites, and other marine, invertebrates, including the strange horseshoe crab (*Limulus*) now found on the east coasts of North America and in Asia. Freshwater forms, on the other hand, are lacking completely. The Solnhofen atolls apparently were situated at some distance from land. Probably the floor of the lagoon was uncovered regularly at ebb tide, while the high tide carried the carcasses into the atoll. As early as 1904 no less than 450 different species of animals had been discovered at Solnhofen.

In a deposit of the same age and type at Sierra de Montsech in northern Spain a great number of exquisitely preserved plant fossils are especially noteworthy, but here are also found frogs and other freshwater animals, sometimes with soft parts preserved; a feather of *Archaeopteryx* has also been found.

Remains of terrestrial reptiles from the Upper Jurassic – sauropods, theropods, stegosaurs, and camptosaurs – are especially common in the Oxford Clay and other deposits in England. However, the majority of the vertebrate fossils belong to marine animals, notably ichthyosaurs, plesiosaurs, and the highly specialised sea crocodiles or thalattosuchians.

Meanwhile, in Tethys the preludes to the Alpine orogeny were going on. Incipient mountain ranges, in the form of island arcs, raised their backs above the sea, while the deep-sea trenches in between were still being filled with sediment.

## The remainder of the world in the Jurassic

Most of North Africa was dry land in the Jurassic, but large tracts in East Africa were flooded. At Tendaguru in Tanganyika a famous fossiliferous series of strata has been found with alternating marine and continental beds, indicating that the level of the sea oscillated several times during a stretch of time extending from the mid-Jurassic to the early Cretaceous. The continental deposits have yielded the bones of dinosaurs that lived in the later Jurassic and show close resemblance to the European and North American ones. The Tendaguru dinosaurs evidently inhabited a low-lying delta, where their fossil bones were imbedded in the masses of silt and mud carried by the sluggish river. At times this type of sedimentation was interrupted by a transgression, which brought in a horizon with marine shellfish. Among the dinosaurs may be noted the giant *Brachiosaurus* and other sauropods, for instance a relative of *Diplodocus*; allosaurs and megalosaurs, small theropods, a stegosaur and a camptosaur. The flying lizards *Pterodactylus* and *Rhamphorhynchus* are also found here, as in the Upper Jurassic of Europe.

East of Europe Tethys extended across the Caucasus and Persia to northern India. Its northern coast lay in South Russia just north of the Caspian Sea during the Lower and Middle Jurassic, but in the Upper Jurassic most of Russia became submerged, and contact was established between Tethys and the Boreal Sea. Many volcanoes were active in the Lias and Middle Jurassic, but this activity was weaker in the Upper Jurassic. The south coast of Tethys extended through the Sinai Peninsula and Arabia.

70 Mesozoic geography of North America. *Left*, uppermost Jurassic, showing invasion by the Sundance Sea from the north, separated from the Pacific Coast geosyncline by the long geanticline of the Mesocordillera. After the regression of the Sundance Sea, the Morrison Formation was laid down at the end of the Jurassic.

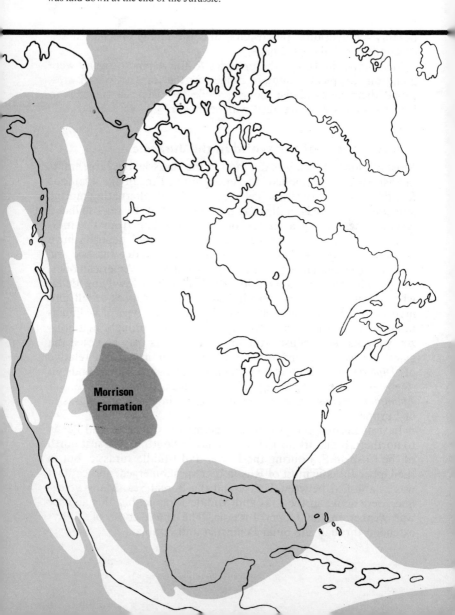

*Right*, Upper Cretaceous transgression. Dashed lines indicate the maximum extent of the interior sea, when the continent was divided into two islands. The situation with separate northern and southern embayments represents the somewhat later time of the Niobrara Sea.

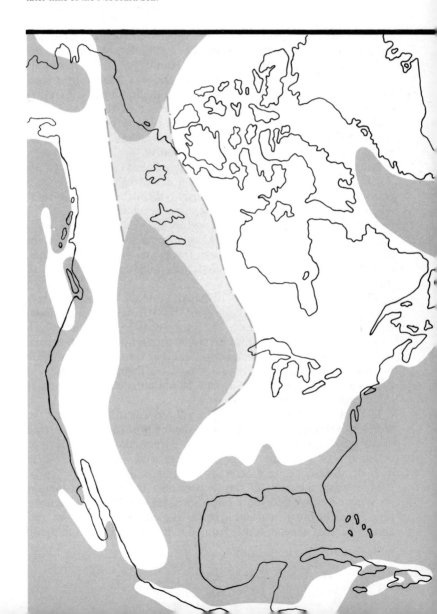

The Indian peninsula, which was mostly emergent, was at this time completely separated from the remainder of Asia by the Tethyan trench along the present-day Himalaya. The western part of the big Indian island was a lowland and was flooded by a shelf sea in the later part of the Jurassic. East of the Indian island, Tethys curved south to embrace Malaya and most of the East Indian archipelago including New Guinea; only the southern part of this island remained outside the Tethyan geosyncline. New Zealand, on the other hand, was linked to the Tethys.

Another branch of the Tethyan geosyncline system passes to the north and through the Philippines, touches eastern Japan and continues by way of Kamchatka right through eastern Siberia, where it comes close to the American geosyncline system.

Australia was mostly emergent, except for shelf seas in the west. Eastern Australia was a low-lying, swampy, wooded area: here are now found coal deposits with fossil freshwater gastropods, plants, and dinosaurs. In East Asia coal beds and dinosaur fossils of Jurassic age are also found.

The Mesozoic geosynclines of North America extended mainly north and south. Most of the present-day Pacific coast area down to California formed a trench, the Pacific coast geosyncline, which was formed in the Trias and persisted throughout the Mesozoic. In the course of the Jurassic period another trench, the Laramide or Rocky Mountain geosyncline, began to form further east, in the area now known as the Rocky Mountains, and both ends of this trench were invaded by the sea.

In the Upper Jurassic an orogenic phase was initiated; now the Sierra Nevada were folded up along the inner border of the Pacific Coast geosyncline. The region now witnessed a great deal of volcanic activity; some of the volcanoes were submarine, others built up their cones to emerge as volcanic islands.

Meanwhile the arms of the sea invading the Rocky Mountain geosyncline converged more and more, yet without ever meeting (fig. 70). The northern embayment, the Sundance Sea, was an arctic sea that spread south from the Boreal Sea and gradually advanced as far as Utah and South Dakota; it was separated from the Pacific Coast geosyncline by an immensely long tongue of land, a geanticline, shielding the cold waters of the Sundance Sea from contact with the warmer Pacific Ocean. In the south the other embayment extended from the Mexican Gulf to cover northeast Mexico and western Texas; but this arm was much shorter than the Sundance Sea.

Towards the end of the Jurassic the Sundance Sea retreated to the north, and its sediments are overlain by the famous, dinosaur-bearing Morrison formation. It consists of river deposits, which conclude the Jurassic System in central North America and cover a wide area from Montana in the north to New Mexico in the south and Utah in the west. This area was then a low plain with sluggish rivers, great swamps and many lakes, with a rich vegetation and animal life. Here lived sauropods of many kinds, from the slender, lengthy *Diplodocus* to the stocky *Brontosaurus* and *Brachiosaurus*; then there were *Antrodemus* and other allosaurs, small theropods, bipedal ornithischians of camptosaur type, stegosaurs, crocodiles, lizards, turtles, frogs, small mammals; it is a swarming multitude of strange beings from the past that we meet with here, in the last few million years of the Jurassic.

The moist conditions in this area contrast with the aridity found in the southwest, in the Colorado Plateau, where one of the greatest dune sand formations known, the Navajo Sandstone, was formed in the earlier Jurassic. Even here, however, fossil tracks and bones of dinosaurs have been found.

In South America a Pacific geosyncline extended along the west coast; Columbia and Venezuela had a dry climate like the Colorado Plateau, with deposition of continental sandstones. Towards the end of the Jurassic there was a phase of mountain building with strong volcanic activity, evidently correlated with the Sierra Nevada revolution in North America. The episode was followed by a transgression. Thus the Jurassic history of South America is rather like that of North America, and the orogeny of the Upper Jurassic seems to have been restricted to these areas.

Otherwise South America was dry land, and the continental deposits yield dinosaur bones. It may thus be said that there were dinosaurs in all parts of the world, and there does not seem to be any marked difference between the types existing in different areas, at least not in the Upper Jurassic. In spite of the high sea level especially in the Liassic, the continuously forming and changing land connections presumably permitted such extensive migration that dinosaurs of all kinds were able to colonise the whole earth.

## The Cretaceous in Europe

The great regression towards the end of the Jurassic led to the extension of the Armorican continent to comprise almost all of England and France, the one-time Brabant Island and part of Central Europe, now emergent to the very border of Tethys. At the same time a shallow trench, the Wealden geosyncline, began to form. In this basin richly fossiliferous delta sediments with remains of land animals were deposited; it was, for instance, in the Wealden beds at Bernissart that the famous Belgian *Iguanodon* herd was found. Here are also found armoured dinosaurs, sauropods, carnivorous dinosaurs and the small, perhaps arboreal *Hypsilophodon*. Crocodiles swam in the river channels, and pterosaurs soared over the delta. Here the fauna of the opening Cretaceous period emerges in all its variety.

Sediments of Wealden type were also formed in other areas, for instance in the Scanian-Danish basin and at the border of Tethys in Spain. The Carpathian arc of Tethys sent a long arm of the sea to the north and west, which passed through northern Germany and Holland to emerge into the North Sea basin and thence to the Boreal Sea.

With the passage from the Jurassic to the Cretaceous, however, once more the seas began to rise. Tethys again flooded its borders, and arms of the sea enveloped the old highlands and drowned the lands of the Wealden dinosaurs. In the mid Lower Cretaceous there followed a minor regression, which was then followed by a fluctuation with a transgressive trend. Towards the end of the Lower Cretaceous, in the Albian stage, a great transgression set in. Now the sea broke through northeastern France and eastern England, joining Tethys once again with the Boreal Sea; the former Brabant-Central European Peninsula now remained as an island, with all its borders encroached upon by the rising waters.

The rule of the reptiles continues on the shrinking mainlands and islands, now in an environment where the old-fashioned bennettitales are yielding to modern flowering plants. Many arboreal mammals hide in the foliage of the hardwood trees, and pterosaurs and birds flutter from tree to tree. Boa-like, large snakes perhaps also wind around the branches; the first snakes appeared in the Lower Cretaceous.

On the borders of the Tethys a new type of sediment called the flysch was deposited, forming beds of fine-grained sandstones and marls many thousand feet thick, which are practically devoid of fossils. Their origin is uncertain, but it has been suggested that the flysch may represent the sedimentation along shelving, mangrove-clad coasts of the type now found in Florida, for instance.

In the sea, the ammonites experience their last great period of prosperity. Here predominates the *Hoplites* type; these are strongly ribbed, more or less flattened, planispiral ammonites. There are also aberrant forms like the turret-like, helicoidal *Turrilites*, and the hook-shaped, partly straight-shelled *Hamites*.

The later Cretaceous is ushered in by a new, immense transgression, and this time almost all of Europe is engulfed by the sea, except a few highlands. Of the Upper Cretaceous Chalk Sea it has been said that it stretched from Texas to the Urals. The Ural Mountains formed an island cordillera in the east. The plateau of Scandis was emergent, but southern Sweden was inundated except for a number of islets and skerries where the Palaeozoic rocks stood out of the water. The Scottish highlands were a half-drowned archipelago; England, Wales, and most of Ireland was under water. Out of this sea world rose the Central European Island and the remains of Armorica: Brittany, the French Central massif, and the Spanish Meseta (see figure 69).

A fairly large island may have existed in the western part of Tethys, comprising Sardinia, Corsica, the Baleares and perhaps part of the present North African coast; it has been thought to have connected with the Central massif by an isthmus, while another peninsula extended along the Pyrenees. North Africa, on the other hand, was almost entirely engulfed.

In the great shelf sea, Chalk was deposited in a broad belt from Ireland to southern Russia. The numerous flints found in the Chalk probably in part were formed from the fossil skeletons of siliceous sponges, which evidently lived in great numbers in this environment. Solitary corals, *Inoceramus* bivalves, *Micraster* urchins and belemnites are other common Chalk Sea forms, while ammonites become rarer in the Late Cretaceous and lose their stratigraphic importance. Many of these free-swimming invertebrates probably provided an important source of food for the sea reptiles, which are varied and striking. Here is found the giant *Mosasaurus*, discovered in the eighteenth century in Belgium, as well as many other mosasaurs. There are also plesiosaurs, both long-necked and short-necked, and forms with medium-long necks. Crocodiles swarmed in the waters; many of them were basically similar to living true crocodiles, and probably only made short excursions in the sea, somewhat like the present-day delta crocodile of southeast Asia. The ichthyosaurs apparently vanished well before the end of the Cretaceous.

The Tethys area shows a division into deep trenches with ammonite-bearing shales, and shallow-water deposits with rudist and coral reefs. The rudist reefs are typical of the later Cretaceous. In other areas belts of flysch and other material derived from the rising island cordilleras are found.

This is the scene in Europe as the sun sets over the world of the dinosaurs. Once more the sea begins to recede, and new land is laid bare. In the Danian strata, which in Denmark and Sweden follow directly upon the Senonian or latest Cretaceous, all the typical Mesozoic reptiles are gone, and the only reptile found here is a gavial-like crocodilian, *Thoracosaurus*. The great reptiles are extinct, and the reign of the mammals is about to begin, all over the world.

## The remainder of the world in the Cretaceous

The Cretaceous sea was transgressive in the other continents as in Europe, and in the late Cretaceous reaches a maximum. The Tethyan geosyncline is of about the same extent as in the Jurassic, from southern Europe over Asia Minor, the Caucasus, Persia, northern India, Malaya and the East Indian archipelago (Borneo, however, was emergent in the early Cretaceous) to New Guinea and New Zealand; and northward via the Philippines and the Japanese east coast to Kamchatka, to connect with the Pacific Coast geosyncline in America. In the Lower Cretaceous, a large part of the interior of Australia became submerged, as well as eastern and northern Siberia as far west as the Ural range. In the Upper Cretaceous a great area lying between Tethys and the Boreal Sea, east of the Caspian Sea and the Urals, was also flooded. Besides, parts of northern India, Borneo, and Japan were inundated in Asia.

The African inundation is still greater. Except for some areas forming islands in the west, the greater part of Africa north of the equator formed the floor of a shelf sea, and the same holds for neighbouring Arabia, and for wide areas along the eastern and southeastern coast of Africa. Madagascar was also submerged in part.

Many of these shallow seas have left fossiliferous deposits. In the Lower Cretaceous of Australia are found plesiosaurs and ichthyosaurs, richly assorted; among the former the fantastic *Kronosaurus* stands out. The terrestrial reptiles have also left numerous fossils, especially in Mongolia, Shantung and Sinkiang. Among earlier Cretaceous dinosaurs may be noted sauropods, theropods, and ornithischians; the Mongolian ornithischians include the interesting little *Psittacosaurus*, which is believed to be ancestral to the Upper Cretaceous horned dinosaurs. Skeletons of the last-mentioned have been found in great numbers in Mongolia, where extensive Upper Cretaceous deposits are known. Upper Cretaceous dinosaurs are also found in China, India, Australia, and Africa; in the latter continent the finds are distributed over the Northwest African islands (Morocco, Sahara) as well as the Nile area.

The Cretaceous history of North America is also characterised by a great transgression, which finally divided the continent into two great islands. After the late Jurassic Morrison regression, the waters started to advance once more in the Rocky Mountain geosyncline. This time the transgression started in the south, forming an embayment from the Mexican Gulf into parts of Mexico and Texas. Next the northern transgression commenced, the Boreal Sea extending into the northern part of the geosyncline. Third, the Atlantic and Gulf coasts were inundated; Florida was completely flooded, forming a shallow underwater bank. Towards the end of the Lower Cretaceous, the Mexican transgression reached Kansas and southeastern Colorado, while the Boreal interior sea entered into northern Colorado. Only a narrow isthmus now separated the two seas, in which the faunas were quite distinct.

Finally, in the beginning of the Upper Cretaceous epoch, came the dramatic moment when the two seas united. North America now formed two islands, and this situation persisted for many million years, while the geosyncline trench sank under the burden of the accumulating deposits, finally forming an immense interior sea a thousand miles wide. At this time about fifty per cent of the continent was under water.

Starting in the north, however, the long trench was gradually filled up again, in the late Upper Cretaceous, with erosion material from the rising highlands in the west; at the same time the orogeny began that was to bring forth the Rocky Mountain range.

After many million years only the southern half of the geosyncline was under water, and then the sea left this part too, until

71 Early Cretaceous seas of Asia, which was isolated from Europe by a seaway connecting the Indian and Arctic Oceans.

at the end of the Cretaceous the entire former geosyncline was dry land. A wide expanse of the area now formed a swampy lowland, over which continental deposits were spread during the last few million years of the Cretaceous. Coal deposits, now found in the youngest Cretaceous strata of the Rockies, were formed in this environment. In many areas the folding and overthrusting of the strata had also commenced, giving rise to many mountain ranges, the first version of the Rocky Mountains; this was the Laramide revolution.

Throughout the Cretaceous, the Rocky Mountain geosyncline was separated from the Pacific Coast geosyncline by a highland area, the Mesocordilleran geanticline; at the height of the transgression this extended as a half-drowned island cordillera from the Arctic Circle into Mexico. All along the range arose active volcanoes, and other volcanoes discharged in the Laramide trench.

In this environment the tremendous drama of reptile history in the Upper Cretaceous was enacted. The fossil record of the Lower Cretaceous is comparatively poor, but as regards the evolution of the fauna in the Upper Cretaceous we are very well informed indeed. This story is told in chapter after chapter, written down in the endless fossil fields of Alberta, Montana, Wyoming, Colorado, Texas, New Mexico, and Mexico, where a series of formations contain dinosaur bones and skeletons. The three main stages in the Upper Cretaceous continental sequence have been named the Belly River, Edmonton, and Lance.

Here we meet with a great fauna in continuous evolution. There are many kinds of duckbill dinosaurs, both crested and crestless; there are bonehead dinosaurs and many kinds of armoured dinosaurs; and great herds of various kinds of horned dinosaurs, which tend to increase in size and finally culminate in the gigantic *Triceratops* and *Torosaurus* of the Lance stage. The sauropods vanish at an early stage in the north but remain to the end of the Cretaceous in Texas and Mexico. Many smaller types of dinosaurs, fast predaceous beasts in most cases, are also known; here belong the ostrich dinosaurs, which managed excellently in their unknown mode of life, and persist into the Lance. The great carnosaurs, now represented by the deinodonts, are less varied but more impressive; here, too, the size culminates in the Lance, with *Tyrannosaurus*.

Besides dinosaurs we know of many other animals that inhabited the Upper Cretaceous woods, lowlands and swamps. Here there were mammals, opossum-like marsupials as well as insectivores side by side with the fruit-eating multituberculates; there were also birds – waders and loons as well as toothed birds. Crocodiles including the colossal *Phobosuchus* of Texas are present in great numbers. We also find lizards, snakes, turtles, frogs, and freshwater fishes.

The interior sea also harboured an impressive fauna, the history of which may similarly be divided into three successive stages – the Benton, Niobrara, and Pierre. Ichthyosaurs are rare and vanish at an early stage, as in Europe. Among plesiosaurs, on the other hand, both the main types – the short-necked pliosaurs with their great jaws and the long-necked, snaky elasmosaurs – are present in undiminished number and variety. Mosasaurs and great sea turtles are other remarkable denizens of the sea. In the same sea lived the swimming bird *Hesperornis*, wingless but with powerful feet adapted for swimming. Many other birds soared over the Upper Cretaceous Niobrara Sea, together with the great, dragon-like pteranodonts. Of course this fauna is not restricted to the interior sea; its remains are also found in other inundated parts of North America.

The geography of the North American east coast was profoundly changed in the Cretaceous, presumably as a result of the continental drift that had removed it from contact with western Africa and Europe. From the Cretaceous on, marine sediments of the coastal plain of eastern North America grade into nonmarine or brackish-water deposits on the landward side, and into finer-grained marine strata towards the sea, But in the older marine deposits of the east there is evidence that the sediments had been derived from a landmass outside the present margin of the continent, and called Appalachia. At the end of the Jurassic this land would seem to have foundered, and so in the Cretaceous the waves of the Atlantic were assaulting the continent of North America for the first time. Indeed, on the drift theory, the Atlantic itself was created in the Jurassic and Cretaceous, when the New World and the Old drifted apart, and the Appalachian and Armorican lands vanished. We may well enquire how these lands were related to each other – Appalachia off the American coast, and Armorica west of Europe. Was Appalachia the southwestern part of Armorica?

In South America the geographic evolution of the Cretaceous was again parallelled by that in North America. An orogeny took place during the Upper Cretaceous in the Andean geosyncline, resulting in the birth of a mountain range extending from Trinidad off Venezuela southwest into Columbia and then southward to and beyond Cape Horn. This corresponds to the Laramide revolution in North America. At the same time a transverse mountain range was folded to form the Antilles system, extending eastward from Central America along the Greater Antilles.

A seaway passed from the Andean geosyncline through Brazil south of the Amazon River to the Atlantic, dividing South America into two islands, both of which were inhabited by dinosaurs. The most important dinosaur faunas have been found in Patagonia and in São Paulo, Brazil.

It may be noted that the history of the Cretaceous in many ways appears to be a cyclic repetition of the Jurassic. In both instances there was a great transgression, succeeded by a regression at the end of the period; there was an American orogeny near the end of each cycle (Sierra Nevada in the Late Jurassic, Laramides in the Late Cretaceous); and the richest land faunas are known from the later part of each period (Morrison in the Jurassic, Belly River-Edmonton-Lance in the Cretaceous).

# 8 After the dinosaurs

## The mass death

The dinosaurs were with us until the very end of the Cretaceous; in our minds we may picture the great sauropods, the herds of *Triceratops*, and the solitary, hungry *Tyrannosaurus*, and feel the ground tremble under their feet. Then we ascend a few inches in the series of strata, and suddenly they are all gone forever, and it is strangely empty around us. No wonder that such a metamorphosis continues to fascinate and attract students.

There are many other times in geological history in which it seems that some kind of mass death has swept away great numbers of forms of life. At the end of the Palaeozoic, and again at the end of the Trias, many kinds of reptiles, amphibians, and invertebrates became extinct. Still, as far as the land faunas are concerned, these changes do not seem as revolutionary, for the new forms that took the place of the extinct were not too unlike their predecessors; as Colbert notes, reptile succeeded reptile. In the transition from the Cretaceous to the Tertiary, the outcome was different: the reptiles that died out were succeeded by mammals and birds.

On land all the dinosaurs became extinct, and their roles were taken over by herbivorous and carnivorous mammals. The flying lizards were succeeded by birds and bats. In the seas the ichthyosaurs and mosasaurs vanished, to be replaced by toothed whales of various kinds, from sperm whales to porpoises. The plesiosaur niche is perhaps occupied by present-day seals. But it is only in the case of the pterosaurs versus the birds that any actual competition could have taken place. In all the other cases the replacing forms evolved long after the end of the Cretaceous.

Not only reptiles became extinct. The toothed birds, for instance, seem to have died out at about this time; the last are found in the late Cretaceous. Invertebrates are also affected by the mass death. The rudists fell victim to it, and the same is true for some gastropods on the same systematic level (superfamilies): Euomphalacea and Trochonematacea, both of which comprise moderately helicoidal shells of rather primitive aspect, and Nerineacea with very high, pointed forms. The ammonites also vanished, while the belemnites became extinct somewhat later.

Many of the invertebrates were also replaced by other forms. The role of the belemnites was taken by the squid. The extinct gastropod groups correspond approximately to other, surviving groups. To find good analogies to the rudists and ammonites may be more difficult; the pearly nautilus comes to mind in the latter case, but it is not a predominant feature in the fauna in the same way as the ammonites.

But of course very far from all the living groups became extinct. On the contrary, there were many, both reptile and others, that survived the transition from the Cretaceous to the Tertiary without decimation. Among reptiles this is true for the turtles, lizards, snakes, crocodilians, champsosaurs, and the tuatara group. As regards birds and mammals, we know that several modern types were present in the Upper Cretaceous. Even an archaic group of mammals like the multituberculates survived into the Tertiary. Looking then at the fishes, we find that the great modernisation set in as early as the Jurassic and Cretaceous, and that the Upper Cretaceous fauna was quite modern in composition. In this way we can take group after group and note that the great changeover was not contemporaneous for all of them.

Exactly what happens when a group of animals becomes extinct – say, a dinosaur order? Does it remain vigorous to the very end, to fall as at a single blow? Or are its ranks thinning for a long time before the final annihilation? When we look back at these events, they are so distant from us in time that it is very difficult to discern such geologically brief intervals as 100,000 or half a million years. And yet these are enormous spans of time, during which a great deal can happen. This means that at present we are unable to observe the details of the extinction at the end of the Cretaceous. Its long prologue, however, can be observed.

Comparing the dinosaur faunas of the Belly River and Lance stages, Colbert found an interesting situation: the Lance fauna, the last of the Cretaceous, tends to be poorer in genera and species that the Belly River fauna, although all of the orders persist. For instance, the ceratopsians or horned dinosaurs, which were represented by 16 genera in the Belly River, only mustered 7 in the Lance. The armoured dinosaurs showed a reduction from 19 to 6 genera, and the duckbills from 29 to 7; only the theropods remained as varied as before (15 genera in the Belly River, 14 in the Lance). Here, then, is a definite trend. Whatever killed the dinosaurs may have begun as a wind of death long before the final catastrophe.

But though the variety in species and genera was reduced, it seems that the number of individuals remained great at the end of the Cretaceous. In the uppermost Lance numerous fossils of the great *Triceratops* are still found, suggesting that large herds of these beasts still roamed the area.

The extinction of the ammonites proceeded in an analogous way. It was preceded by a long-term reduction of the number of species, so that only a few forms remained at the end of the period.

Let us make a chronological summing up of the process:

Lower Cretaceous: Stegosaurs, thalattosuchians, bennettitales become extinct; holosteans are reduced in number. Armoured dinosaurs and snakes appear, and the same seems to be true for the flowering plants, the champsosaurs and the modern toothless birds. Teleosts increase in number.

Upper Cretaceous: Pterosaurs, ichthyosaurs, and probably toothed birds became extinct. Ceratopsians arise and develop greatly, but are again reduced towards the end of the period. Ornithopods, armoured dinosaurs, ammonites, and cycads are also reduced. New important groups are marsupials, placentals, and rudists. An increase may be seen among modern birds, modern crocodilians, modern sharks and rays, teleosts and flowering plants.

Transition Cretaceous-Tertiary: Theropods, sauropods, ornithopods, ankylosaurs, ceratopsians, plesiosaurs, mosasaurs, ammonites and rudists become extinct. Amphichelydians and belemnites are reduced in number.

In this perspective the extinction at the end of the Cretaceous becomes the culmination of a process that had been going on for a long time but was gradually intensified. The main difference is that the extinctions earlier in the Cretaceous tended to be balanced by the rise of replacing groups: there was a give-and-take between death and renewal, debit and credit. At the end of the Cretaceaus, on the other hand, a great number of forms died out without being replaced by others, so that a great impoverishment of the fauna was the consequence.

The Age of Reptiles was over; the Age of Mammals was beginning. Let us now take a look at the Cenozoic era, the last and the shortest of the eras in the history of the earth. Finally, with this added perspective, we may return to the problem of extinction.

## The Palaeogene

The Cenozoic is divided into two periods (Tertiary and Quaternary) and seven epochs, which have recently been potassium-argon dated as shown in the diagram. The Tertiary epochs may be grouped into the Palaeogene or older Tertiary (with the Paleocene, Eocene, and Oligocene epochs) and the Neogene or newer Tertiary (Miocene and Pliocene). The boundary is somewhat artificial, but the Palaeogene and Neogene when viewed in their entirety are very different as regards fauna and geography. The Palaeogene would seem to us an alien world, populated by completely strange and unknown animals, and almost unrecognisable geographically. The Neogene, on the other hand, has a fauna that in part, at least, would have a familiar look, and some basic land forms that we might recognise – such as the early version of the Alps.

The geography of Europe in the early Tertiary was greatly affected by the earlier phases of the Alpine revolution, in which the Alps and other young mountain ranges began to emerge from the Tethyan deep in the form of island cordilleras. The Pyrenees and Apennines were formed mainly in the late Eocene, while the Alps themselves underwent a major folding and uplift in the Oligocene. Meanwhile, shallow seas advanced and retreated over the low-

**Table 3** Earth history in the Cenozoic era

| Period | Epoch | Age in millions of years | Important mammals |
|---|---|---|---|
| Quaternary | Holocene | 0 – 0.01 | Man, domestic animals |
| | Pleistocene | 0.01 – 3 | Man; ape-men; elephants; bovines |
| **Tertiary** | | | |
| Neogene | Pliocene | 3 – 12 | Three-toed steppe horse; antelopes; mastodonts |
| | Miocene | 12 – 25 | Mastodonts; rhinoceroses; forest horses |
| Palaeogene | Oligocene | 25 – 36 | Oreodonts (N. America), rhinoceroses |
| | Eocene | 36 – 55 | Creodonts, titanotheres, Dinocerata |
| | Paleocene | 55 – 63 | Archaic mammals |

lying plains, but there was never a transgression of the magnitude attained by the Chalk Sea, and most of northern and central Europe remained emergent. At the end of the Palaeogene, in Oligocene times, the last great transgression of the era inundated much of Europe and created a seaway from the English Channel to Russia across the Netherlands, Germany, and Poland.

The European climate was generally warm in the Palaeogene, though there may have been a slightly cooler episode at the very beginning of the Cenozoic; it is, however, insufficiently documented. Perhaps the cool facies of the London and Paris basin deposits from the Paleocene epoch is simply due to the fact that these marine embayments communicated directly with the Arctic by way of the North Sea, while there was land in the west – a remainder of Armorica, or a land bridge to North America. In the Eocene the land was severed by a marine strait, so that warm Atlantic water entered the London and Paris basins. It also brought here the typical Palaeogene giant foraminifers, the nummulites – flat, coin-like shells up to four inches in diameter. A related present-day form is found in shallow tropical seas, and so it would seem that both the Tethys and the Atlantic of the Palaeogene were

72 Extreme fragmentation of land areas is characteristic of the early Tertiary. Like so many Noah's Arks, each fragment tended to develop and support a unique land fauna of its own. Gradually during the Cenozoic the fragments united to form larger blocks, and the land animals tended to disperse, a process which is still going on. This is the historical background to the great variety and persistent local differentiation of the modern land fauna.

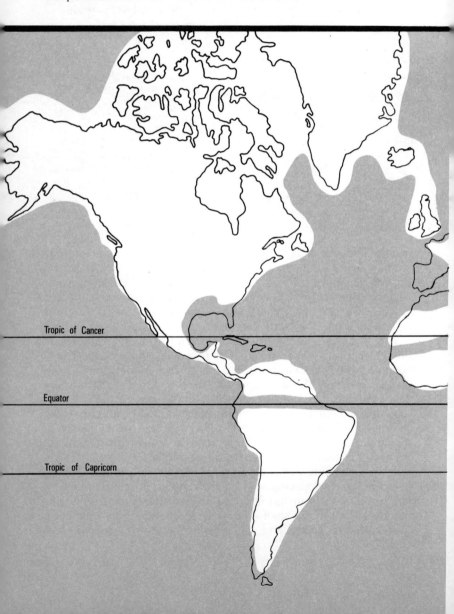

warm. The flora of the Eocene London Clay also testifies to the warm climate, for it is a jungle flora resembling that of Malaya today.

Africa was still divided into three separate islands in the Eocene, and much of North Africa was under a shelf sea. The Palaeogene mammals of this continent clearly show the results of the isolation, for they are quite different from those of Eurasia and North America.

East of the Ural mountains a seaway extended from Tethys to the Arctic Ocean, and completely severed the connexion between Europe and Asia in the earlier Palaeogene. A regression in the middle Eocene finally opened up a land bridge across this seaway. The land mammals of the two continents, which up to then had been quite dissimilar, were now able to migrate across the bridge. From this time on, Europe and Asia have remained in contact, and form a single faunal province, the Palaearctic.

North America was connected with Asia by way of a bridge across the Bering strait, either intermittently or constantly during the Palaeogene. Since the regression at the end of the Cretaceous there has been no great inundation of the North American continent. Florida and much of the Gulf Coast, however, were still under water in Palaeogene times, and part of the Atlantic coastal plain was also inundated. The Appalachians were largely worn down to a flat surface (a peneplane) in the Palaeogene; later on, the whole region was uplifted, and renewed erosion cut new valleys and so brought forth a new version of the mountains.

Something like this also occurred in the Rocky Mountains, which were formed in the Upper Cretaceous Laramide Revolution. In the Palaeogene, the intermontane valleys were gradually filled with sediments, while the mountains were worn down. Remains of these sediments have yielded an extraordinary assemblage of fossil mammals from the Paleocene and Eocene. Later on, the region was uplifted, and there followed dissection in the course of which the ancient mountains were largely cleared of the debris in which they had been hidden.

Much of the west coast geosyncline was still under water in the Palaeogene, and the coast lines in California were highly variable. At its maximum the transgressive sea lapped at the foothills of the Sierra Nevada, while the first anticlines of the Coast Range emerged as islands.

North America seems to have been in direct contact with Europe by a land bridge in the early Palaeogene; the bridge probably extended from the British Isles across Iceland and Greenland. Faunal exchange was evidently very intense, and in the early Eocene the resemblance between the European and North American mammals was so close that the two continents essentially formed a single faunal province. However, by mid-Eocene times the land bridge would seem to have foundered, and from this time on the resemblance rapidly diminished. Thenceforth migration between Eurasia (the Palaearctic) and North America (the Nearctic) seems to have occurred by the Bering route only.

Judged from the history of its land mammals, South America became completely isolated from all other continents at the beginning of the Cenozoic, and remained an island continent almost to the end of the Tertiary. A rich series of fossil deposits in Patagonia gives a remarkable record of the Palaeogene mammal life of this continent, first studied by the great pioneers, Florentino Ameghino and his brother Carlos, in the late nineteenth century.

The Palaeogene history of Australia is little known, but presumably this continent was a stronghold of the marsupial mammals throughout the Cenozoic. Whether placental mammals were present in Australia in the Palaeogene is not known; but if so, they must have become extinct locally, for all the placental mammals now present here are comparatively late invaders.

Let us now take a look at the animal life as it unfolded in the Palaeogene, after the disappearance of almost all of the great reptiles of the Mesozoic. In the beginning of the Paleocene epoch, the largest land animals were still to be found among the reptiles; there were no mammals to rival the crocodiles, the champsosaurs, and the big python-like snakes in size. The champsosaurs became

73 Some dominant reptiles of the Age of Dinosaurs, compared with ecological counterparts among living animals. The similarities have arisen, not through genetic relationship, but from adaptation to a similar mode of life, illustrating the evolutionary process of convergence. The soaring reptile *Pteranodon* resembles a modern albatross. The similarity between the horned dinosaur *Triceratops* and a rhinoceros is also obvious, whereas the sauropod *Brontosaurus* and a modern elephant, both with the role of terrestrial giants, are not much alike. Again, the predaceous

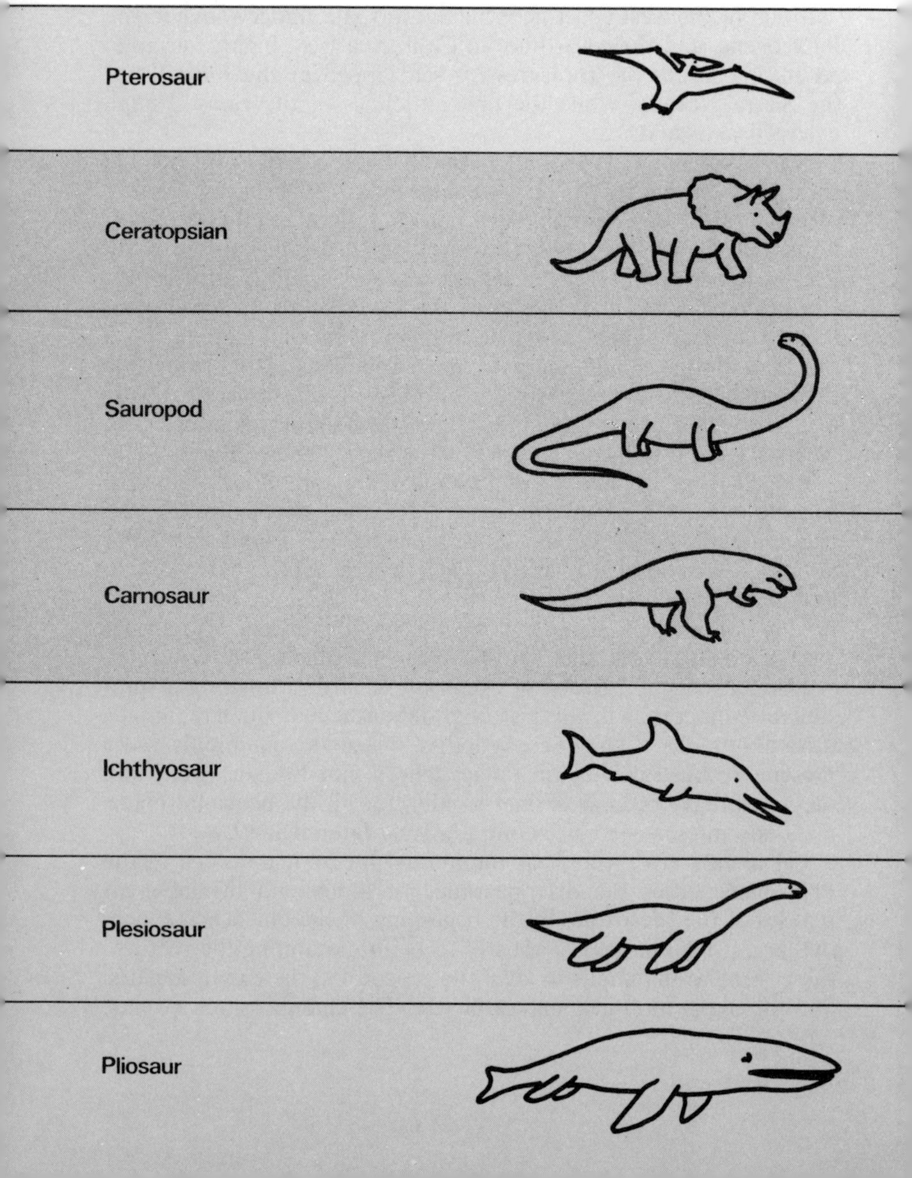

Pterosaur

Ceratopsian

Sauropod

Carnosaur

Ichthyosaur

Plesiosaur

Pliosaur

*Megalosaurus*, though ecologically corresponding to the great cat, does not resemble it morphologically. In marine forms, the resemblance is much closer, as between *Ichthyosaurus* and a dolphin, between *Plesiosaurus* and a seal, and between *Kronosaurus* and a sperm whale. Evolutionary convergence, while creating resemblance, can never lead to identity, partly because of different heritages of the groups, partly because ecological roles never repeat exactly; history does *not* repeat itself.

| | |
|---|---|
| | Bird |
| | Rhinoceros |
| | Elephant |
| | Great cat |
| | Dolphin |
| | Sea lion |
| | Whale |

extinct in the Eocene, but the other reptile groups flourished, and tropical climates permitted the continued existence of crocodilians and other warmth-loving forms in high latitudes.

The reptiles that survived into the Tertiary were already highly adapted to their special modes of life, and they did not change their ways of life so as to take over the roles of the dinosaurs as the dominant terrestrial herbivores and carnivores. Among the birds, however, there was a tendency to evolve great, flightless, rapacious forms. The Diatrymiformes were such a group of great crane-like birds up to two metres tall, which terrorised Europe and North America in the Paleocene and Eocene. Independently, another crane-like group (the cariamas) gave rise to an Oligocene South American family of flightless birds, the Phororhacidae; the largest may have measured three metres in height. The related contemporary thunderbirds (Brontornithidae) were even more heavily built. Both groups persisted in the Miocene.

Most of the present-day orders of birds were well established in the Palaeogene, and many of them date back to the Cretaceous. On the other hand, the toothed birds of the Mesozoic were extinct. In an extinct order of large, shore-living birds (Odontopterygiformes) there were bony, tooth-like spikes in the jaws. This group of pseudo-toothed birds survived into the Neogene.

The mammals of the Paleocene epoch represent several orders, most of which probably evolved in the Cretaceous. Some of the orders still exist, while many others became extinct well before the end of the Oligocene, perhaps as a result of competition from more advanced forms. Among these were the Pantodonta, the first mammals to reach giant size, culminating in the big, hippopotamus-like *Coryphodon* so common in the early Eocene of Europe and North America; the very ungainly Dinocerata, with forms like *Uintatherium* of the Eocene, with big sabre-like tusks and three pairs of horns; and the Condylarthra or primitive hoofed mammals. The last-mentioned, though extinct as such, still survive in the daughter orders Perissodactyla and Artiodactyla, the modern hoofed mammals (ungulates), both of which arose in the Eocene.

On the other hand, many mammalian orders from the Paleocene epoch are still with us. Perhaps the most interesting are the Primates, which were then represented by lemur- and tarsier-like forms, some of them with rodent-like teeth. On the other hand, some Eocene lemurs are clearly close to the main stem leading to the rise of higher primates. In the early Oligocene, Africa was inhabited by primitive ape-like forms. Some of these, notably a small creature called *Propliopithecus*, in which the canine teeth were very small, may even be on the line leading to man.

Other Paleocene orders of modern type are the insectivores, the carnivores, the sirenians (sea-cows), the edentates, the rodents, and the lagomorphs (hares and rabbits). The most conspicuous carnivores of the earlier Palaeogene were the creodonts, which lack the specialised carnassial cheek teeth typical of modern carnivores. Later on, they were replaced by representatives of modern families, including sabre-toothed cats and various types of dogs. The rodents and lagomorphs introduced an element of competition that was to prove fatal to the long-lived Multituberculata. This ancient order, which we have already met in the Mesozoic, became finally extinct in the Eocene.

The main innovations of the Eocene epoch are the two ungulate orders already mentioned, together with the whales (order Cetacea), the elephant order (Proboscidea), and the bats (Chiroptera). The odd-toed and even-toed ungulates have a very good fossil record. The former (order Perissodactyla) produced an impressive array of large forms such as the rhinoceroses, the titanotheres (superficially rhinoceros-like forms that tended to develop a Y-shaped nose horn), the chalicotheres (which were armed with enormous digging claws instead of hoofs), and various early, horse-like types. A hornless rhinoceros of the Oligocene in Eurasia, *Indricotherium*, was the largest land mammal that has ever existed. The even-toed ungulates were also highly varied in the Palaeogene; most types were somewhat piglike, but there were also primitive camels.

The first proboscideans arose in Africa, then an isolated continent, and were confined here to the end of the Palaeogene.

Meanwhile, the island continent of South America became populated by a mammalian fauna, in which basic types similar to those in the other continents evolved quite independently. The carnivorous niche was occupied by marsupials, which radiated into many adaptive types resembling dogs, martens, and sabre-toothed cats. The placental herbivores, on the other hand, evolved to produce mammals resembling horses, camels, elephants, hippopotami, as well as some unique and weird forms like the astrapotheres – big animals with powerful tusks but peculiarly feeble feet and hind quarters. The edentates, originally present in both North and South America, were soon confined to the latter continent, where they radiated to give rise to armadillos, great armoured glyptodonts, ant-eaters, tree sloths, and giant ground sloths.

In the Oligocene, a group of rodents reached South America, probably as a result of accidental rafting. They gave rise to the American porcupines and the diverse South American rodents (chinchillas, guinea pigs, capybaras, coipu, etc.).

This splitting up into isolated continents, each of which becomes populated with a distinct fauna, is in decided contrast with the situation in the Mesozoic, when dinosaurs of very much the same type were found in all parts of the world. The isolation was probably produced in part by continental drift, and in part by the great Cretaceous transgression.

The seas were also invaded by mammals: by the sea-cows in the Paleocene, and the whales in the Eocene. While the former are probably related to the elephants, and so perhaps are of African origin, the whales may have arisen from early carnivores. The Eocene whales were great serpent-like creatures, which probably resembled their ecological predecessors, the mosasaurs, both in external appearance and in mode of life. Ray-finned fishes and sharks continued to dominate the marine fish fauna, and one of the species of large white shark (*Carcharodon*) is represented by fossil teeth indicating a mouth gape of about two metres! Among invertebrates, nummulites are so common in the European marine

74 Palaeogene mammals. (a) *Barylambda*, a large Paleocene mammal with an almost reptile-like tail; (b) *Mesonyx*, an Eocene creodont (primitive carnivore); (c) *Indricotherium*, an Oligocene giant hornless rhinoceros; (d) *Brontotherium*, an Oligocene titanothere; (e) *Uintatherium*, an archaic Eocene mammal; (f) *Archaeotherium*, a giant pig-like animal from the Oligocene.

deposits that the Paleocene and Eocene are sometimes called the Nummulitic. In North America, however, their distribution is more restricted. The belemnites, which had been so common in the Cretaceous, survived into the early Palaeogene, then followed the ammonites into oblivion.

## The Neogene

In the Neogene, the waters gradually receded from the basins that had been flooded in Europe, while the Alpine orogeny proceeded in several phases during the Miocene and Pliocene. Orogenic activity was also pronounced in Asia, North Africa, and both Americas. The regression of the seas now resulted in land connexions between Africa and Eurasia, and caused the former continent to emerge completely; from now on, faunal interchange between Africa and the Palaearctic was intense. The Bering bridge appears to have functioned intermittently. In the area of Central America, various large islands provided stepping stones for a number of land animals that migrated between North and South America, and by mid-Pliocene times a complete isthmus seems to have formed; from then on, migration was intense. Australia, on the other hand, remained isolated.

The climate of the Neogene fluctuated, but a cooling trend is apparent. In the early Pliocene, rainfall was markedly reduced in the temperate areas, and as a result, great grass steppes and savannas were formed; they evidently carried an animal life rivalling that of the present-day African savannas in richness and diversity. The gradual cooling of the climate led to the disappearance of tropical plants and animals in the higher latitudes; the last crocodilians in Europe, for instance, date from the Miocene.

The land faunas show a marked modernisation as compared with the Oligocene, mostly because of the extinction of a large number of ancient mammal families. The early lemur- and tarsier-like primates were gone, and the same is true of nearly all the creodonts (or early Carnivores), as well as a great number of archaic

75 Neogene mammals.
(a) *Platybelodon*, a Miocene shovel-tusked mastodont;
(b) *Syndyoceras*, a Miocene ungulate related to the chevrotains;
(c) *Moropus*, a clawed ungulate or chalicothere from the Miocene;
(d) *Teleoceras*, a Mio-Pliocene short-legged, amphibious rhinoceros;
(e) *Hipparion*, a three-toed Pliocene and Pleistocene horse.

ungulates. Instead, the first deer and antelopes appeared in the Old World; the former soon became numerous in the forested areas, whereas the latter prospered in immense herds on the grasslands. In North America their place was taken by the pronghorn group and also by the various types of horse, for the horse family had had its evolutionary centre in North America since the Eocene. Here, the three-toed forest horses of the Palaeogene gave rise to three-toed steppe horses in the Miocene, and these finally were succeeded by one-toed steppe horses in the Pliocene. Successive migrants of these horses are seen in the Old World: three-toed browsing *Anchitherium* in the Miocene, three-toed grazing *Hipparion* in the Pliocene. Finally, the advent of the one-toed true horse, *Equus*, in the Old World is taken as the time marker for the beginning of the Quaternary.

The Neogene might almost be termed the Age of Mastodonts, for these proboscideans now spread from Africa into all the northern continents and finally also into South America, evolving into numerous remarkable forms.

All the modern carnivore families are present in the Neogene, when bears and hyenas make their first appearance. The cats were still mainly represented by sabre-toothed types; the dog family comprised not only wolf-like and fox-like forms but also big bear-dogs, hyena-like dogs, and cat-like dogs.

Among primates, both apes and monkeys became widespread in the Neogene. The former were medium-sized to large species of *Dryopithecus*, related to the chimpanzee and gorilla. In the Miocene the earliest known hominid, *Ramapithecus*, enters the record. It is found both in Africa and Asia, and probably also in Europe. It was a small form, perhaps the size of a child of five, but unfortunately we know little more than the dentition; this, however, is closely similar to that of the earliest Quaternary hominids. The ancestry of man may thus be traced back some 20 million years, which is the approximate date of the earliest known *Ramapithecus*.

With the emergence of the land bridge faunal migration between

South and North America started and finally led to the extinction of many of the typical indigenous South American faunal elements. The first immigrants to appear in the Neogene are the Miocene monkeys of South America, apparently derived from North American Oligocene lemurs. They were probably accidentally rafted from island to island, like the rodents in the Oligocene. In the Pliocene a few South American forms, including armadillos, glyptodonts, ground sloths, and porcupines, entered North America; but the main part of the migration went in the other direction and brought into South America a host of ungulates and carnivores that were gradually to oust the original inhabitants.

Australia remained in isolation. The rather sparse fossil record of the Neogene shows the presence of a varied marsupial fauna with many interesting forms, some of which grew to a very large size.

The whales and dolphins of the Neogene seas were essentially modern in type. Walrusses, seals and sea lions were present from the Miocene on; they have evolved from terrestrial carnivores.

## The Quaternary

The gradual cooling of the climate led finally to the formation of great continental ice sheets, and so brought about the Pleistocene Ice Age. We have to go back to the Carboniferous and Permian, some 250–300 million years ago, to find the previous ice age. A still earlier ice age took place just before the Cambrian, or some 600 million years ago. The Ice Ages thus seem to have a periodicity of about 300 million years.

The Pleistocene Ice Age, however, was not a uniformly cold epoch. On the contrary, glaciations alternated with mild intervals or interglacials, when the climate was approximately the same as now. Then the great ice sheets melted away, leaving only the Antarctic and Arctic ice caps. There is evidence of a similar alternation in the Carboniferous Ice Age, for instance in the rhythmic pattern of transgressions and regression of the Coal Measures. In the same manner, the Pleistocene seas rose during interglacials, and receded in the glacial ages, when a great volume of water was bound up in the land ice.

Exactly how long the Carboniferous Ice Age lasted is unknown, but it probably went on for many million years. There is no reason to assume that the Quaternary Ice Age will be shorter. So far it has only lasted about a million years, which suggests that we are now probably in an interglacial, and that glacial conditions will return in future.

Within the last million years at least four major glacial ages have taken place. In the Northern hemisphere their record is found in the form of deposits laid down by the ice, as well as in various other kinds of fossil evidence of a cold climate. Such a great extension of the Arctic biotope brought about an unprecedented evolution of animals and plants adapted to cold conditions. Each glaciation brought a flora and fauna of this type into the temperate areas of the present day, while the normal temperate-climate fauna and flora retreated to the south, where many of the desert areas of today (Sahara, southwestern United States) were well watered at the time.

As we have seen above, the Pleistocene epoch, and hence the Quaternary period, is now usually regarded as beginning with the appearance in the Old World of one-toed horses (*Equus*). To them may be added true elephants (as opposed to mastodonts) and cattle-like bovids. On this basis the Pleistocene began about 3 million years ago, judging from the radiometric datings of the earliest deposits with such guide fossils. The first two million years of Pleistocene time were thus preglacial, although climatic fluctuations of a less intense type were going on throughout this time.

The mammalian fauna of the present day was already in existence in the Pleistocene. In addition there were a great number of forms now extinct, many of them of large size. The true elephants (which differ from mastodonts in the more advanced structure of their teeth) now proliferated mightily, producing three main lines of evolution: the savanna elephants of Africa, the forest elephants of Asia and Europe, and the steppe elephants or mammoths of

76 Pleistocene mammals. (a) *Megaloceros*, giant deer or
Irish elk; (b) *Smilodon*, a large sabre-toothed cat;
(c) *Ursus spelaeus*, the great European cave bear;
(d) *Mammuthus primigenius*, the woolly mammoth; (e) *Doedicurus*,
a glyptodont, or giant form related to the armadillos; the
ultimate size in mammals tends to come in the Pleistocene.

the northern steppes and tundras. In many areas, mastodonts and other less progressive proboscideans persisted together with the elephants.

There were giant deer in the Pleistocene, also giant pigs, giant oxen, and giant buffaloes. There were even giant lemurs in Madagascar. The carnivores produced giant cats and bears. A giant beaver the size of a black bear is found in North America. The ground sloths and glyptodonts, found in both Americas, also produced giant forms; excavated by Darwin in Argentina, these helped to impress him with the idea of organic evolution.

The rhinoceroses flourished in the Old World with such forms as the woolly rhinoceros and the elephantine *Elasmotherium*, which carried a single frontal horn two meters in length. Their place in the North American Pleistocene was occupied by the bisons, Old World immigrants that proliferated and spread all over the continent. The endemic South American mammals also produced some very large forms, and in isolated Australia there were giant kangaroos and wombats. The great majority of the Pleistocene giants became extinct without issue; some survived in a dwarfed condition.

The hominid family, of which we got our first glimpse with *Ramapithecus* of the Neogene, reappears in the early Pleistocene with *Australopithecus* in Africa; the earliest representatives are dated at almost 2 million years. They were upright, bipedal forms with a dentition of human type, but with very small brains, hardly exceeding those of the larger apes in size. *Australopithecus* was able to shape simple stone tools, and hunted all kinds of game for food. There is also a related, more brutish-looking form called *Paranthropus*, which some authorities regard as a more vegetarian, forest-living hominid. This ape-man had very large cheek teeth, presumably to cope with a highly abrasive diet.

More nearly human forms appear at a somewhat later stage; this is the *Pithecanthropus* (or *Homo erectus*) group, which was in existence at least 1 million years ago, and persisted up to perhaps 200,000 years ago in some areas. Hominids on this level have been found in Africa, Java, China, and Europe. This group of early humans invented the use of fire, and developed stone tools and weapons of fairly high quality.

The later Pleistocene is the time of Neanderthal Man and modern man (*Homo sapiens*). Their exact relationship to each other and to the pithecanthropine type, which appears to have persisted in marginal areas at the same time, is much disputed. In Europe a Neanderthaloid line of evolution may be traced from the First Interglacial (Heidelberg Man) through the Swanscombe and Steinheim types of the Second Interglacial, and up to the true Neandertalers of the Third Interglacial and Last Glaciation. The modern type of man appeared in Europe some 30,000 years ago, and replaced the Neanderthalers. Some authors think that modern man evolved in this area from Neanderthal ancestors, while others regard him as an immigrant. The latter interpretation seems somewhat more probable, considering the relatively sharp shift from the Neanderthal type to the modern form, and also seeing that some extra-European finds of modern man(e.g. in East Africa and in Sarawak) seem to be older than the first record in Europe. Neanderthal flint industries show a high degree of workmanship, yet are completely eclipsed by the late Palaeolithic industries ascribed to modern man, in which the range of variation and the craftsmanship may be truly remarkable.

Modern man was certainly present in North America at the end of the Pleistocene, but the exact time of his immigration is not yet certain. There is some evidence that America may have been invaded at about the same time as Europe.

The post-glacial or Holocene epoch spans only the last ten thousand years or so. In this time falls the entire evolution of human culture from the Palaeolithic: the invention of agriculture and animal husbandry; the development of urban civilisation; the discovery of metals; the technical revolution.

The late Pleistocene and Holocene was, like the end of the Cretaceous, a time of great extinctions. The parallel is quite close because, as in the Cretaceous, there has been no immediate replacement; the fauna is simply becoming impoverished. This process is still going on, and is now almost entirely due to the activity of man. Human agency probably also was important in the earlier phases of the extinction, but the great climatic changes in the late Pleistocene may also have had an important effect.

On the other hand, the Pleistocene extinction does not have that rather uncanny aspect of wholesale destruction that is so typical of the Cretaceous. Among the mammals, only two orders became entirely extinct at the end of the Pleistocene, and they were the endemic South American orders Litopterna and Notoungulata, which were of local importance only. In contrast, the Cretaceous extinction wiped out most of the dominant orders of reptiles all over the world.

It is of course possible that the same fate will in the near future befall many of the mammalian orders. The orders Dermoptera, Tubulidentata, Sirenia, Hyracoidea, and Proboscidea contain only one or a few species each, and might easily be eradicated. But among these only the proboscideans have ever played a dominant role in the fauna. To produce a real parallel to the Mesozoic extinction, we should at least have to envisage the passing, in addition, of the odd-toed and even-toed ungulates, the carnivores (including seals), the bats, and the whales and dolphins!

## Why did the dinosaurs become extinct?

Is it possible to explain the processes of extinction, as manifested in the mass death of the dinosaurs? G.L.Jepsen recently made an inventory of the hypotheses that have been put forth (in a more or less sincere frame of mind). A list might read somewhat as follows:
1. Climatic change: the dinosaurs died of cold or heat, drought or excessive moisture.
2. Food problems: the dinosaurs did not get enough food, or perhaps ate themselves to death; there was a lack of important minerals, perhaps due to drought; the water may have been poisoned by toxic substances.

3 Disease, parasites, or internecine fights killed the dinosaurs.
4 The dinosaurs were badly constructed anatomically or physiologically, and so were weakened by slipped vertebral discs or hormonal disturbances.
5 The composition or pressure of the atmosphere was changed, perhaps because of volcanic gases, volcanic ash, meteorites, comets, or too high oxygen production by the green plants.
6 Certain mammals specialised in eating dinosaur eggs and thus extinguished the dinosaurs.
7 The dinosaurs were an old race, which became senile and died.
8 The carnosaurs became so efficient that they exterminated their prey (the herbivorous dinosaurs) and died of hunger.
9 The cosmic radiation increased greatly, perhaps as a result of the explosion of a supernova, and produced fatal mutations.
10 The moon was pulled out from the Pacific, which resulted in earthquakes and tidal waves, killing off all the dinosaurs.
11 Mountain building, shifts of the pole, changes in the rotation of the earth, draining of the swamps, floods – all of these might have ended the story of the dinosaurs. They have also been thought to have died because of the feebleness of their brains.
12 God's will, suicidal psychosis, raids by little green men in flying saucers, lack of space in Noah's ark, and palaeoweltschmerz, are also among the suggestions.

Looking at this list one may be surprised that the dinosaurs managed to hold out as long as they did.

Some of the suggestions are not really worth analysis, but others are plausible enough to merit discussion. It may then be noted that great catastrophes of various kinds have been suggested: floods, earthquakes, volcanic eruptions producing deadly gas and ash falls, and the like. This may sound reasonable enough, yet one must also remember that many other forms of life would have passed through unscathed, although they should also have been drowned or poisoned or grilled to death. Again, a flood might have killed many land animals, but would not have affected marine forms like the plesiosaurs and mosasaurs. Furthermore, the catastrophe would have had to be almost universal in proportions, as we know that dinosaurs were present in most or all continents. Such catastrophes would surely have left some geological evidence. But the transition from the Cretaceous to the Tertiary, in for instance the famous succession in Alberta, shows nothing of the kind; the strata continued to pile up peacefully, only there were no more dinosaur bones embedded.

As to the ripping out of the moon from the basin of the Pacific, this theory still finds support among scientists. If this occurred, however, it probably happened at a much earlier stage, for even as far back as in the Devonian there was a moon and a 30-day month, as shown by the tidal effects on the daily growth of Devonian corals.

There are also objections to the more subtle theory of catastrophe, that which postulates an increase in radiation, producing fatal genetic effects. Why should only these special groups have been affected, while others were not?

It has been remarked that the great reptiles, because of their high degree of specialisation, would have been especially vulnerable to mutations. Indeed it can hardly be doubted that some giant sauropods were close to the limits of mechanical efficiency. Still I do not see how you could validly claim that a mosasaur, for instance, was more highly specialised than, say, a sea turtle, a bird, or a mammal.

Food problems may also have been real. The vegetation underwent great changes in the course of the Cretaceous. If some of the dinosaurs were specially adapted to feeding on some kind of plant and unable to switch to other sources of food, the extinction or scarcity of that special plant would have affected them. The present-day koala or marsupial 'bear' will only eat eucalyptus leaves, and cannot survive in an environment where this tree is lacking. However, it does not seem very likely that the dinosaurs were so narrowly specialised, at least not the more widely distributed forms.

A variation of these biological trends of thought is seen in the popular idea that the mammals might have been exterminated the dinosaurs by eating their eggs. It is, of course, not in the least improbable that some mammals (as well as some reptiles and birds) did eat dinosaur eggs, but this is not likely to lead to the extinction of dinosaurs. The present-day Nile monitor, for instance, avidly hunts and eats the eggs of the Nile crocodile, without however exterminating its larger relative. And, of course, egg-hunters would not have affected viviparous marine reptiles.

As regards parasites and diseases, it may be thought that one or a few species could succumb; but an epidemic simultaneously affecting so many different forms in such highly varied environments does not seem credible.

Thus, among the many hypotheses, some may appear plausible enough, but there is no positive evidence for any that have been discussed so far. Faced with all these blind alleys, some speculative authors have postulated that the extinction of the great reptiles did not result from external factors, but from internal. Orthogenesis with resulting over-specialisation was one such suggestion; it has already been discussed and dismissed in connection with the coiled oysters, *Gryphaea*.

Another suggestion emerged from the comparison of animal populations with individual beings, which are born, have a period of growth, a mature age, and finally a period of senility and eventual death. Both the dinosaurs and the ammonites have been cited as examples of a 'senile' race. The appearance of aberrant shell forms in certain Cretaceous ammonites was seen as a symptom of racial senility, like the 'overspecialisation' of the dinosaurs.

For this phenomenon in the ammonites, the explanation is probably simple enough, as shown by G. G. Simpson. If an animal group becomes subjected to greater environmental pressure than before, natural selection will tend to favour variants from the normal type. In such a situation, a reduction in numbers, as well as the appearance of unusual variants, is to be expected. This agrees with the trend seen in ammonites. Aberrant shell form is thus no symptom of senility, but on the contrary may be regarded as a result of undiminished vigour and evolutionary potential in an increasingly hostile environment. Besides, it is obviously incorrect to compare the life of a species or phylum with that of an individual, since the species is constantly being renewed by the birth of new individuals. Again, why has 'racial old age' not killed off such prize oldtimers as the brachiopods *Lingula* and *Crania*, for example?

Speaking of the dinosaurs, a little reflection should make it clear that we can hardly use the phrase 'over-specialised' about animal types that existed for more than a hundred million years without further change, as in the case of the sauropods. Also, it can hardly be said that the dinosaurs that lived in the Lance age were notably more specialised than their predecessors in the Belly River age. And the idea that dinosaurs were 'badly constructed' is also negated by their long term of existence.

A more reasonable explanation might then be a change in climate, perhaps connected with the large-scale orogeny that was initiated in the Upper Cretaceous, and with the accompanying great regression. It is interesting that the Danian (basal Tertiary) is regressive almost everywhere; the Scanian-Danish basin is exceptional in this respect. A combination of these factors may have led to a certain cooling of the climate, which would of course affect the reptiles to a greater extent than the birds and mammals, which can regulate their body temperature. It is striking that all of the typical Mesozoic reptiles are gone in the Danian.

There is in fact some independent evidence supporting this idea – which is more than can be said about other attempted explanations. The studies of dinosaur eggs by the French scientists have given some remarkable information. In the uppermost Cretaceous certain pathological features in the microstructure of the egg shells appear to become more common. It looks as if the formation of the shell was interrupted for some time, to commence again after an interval. This may suggest that the reptiles were occasionally chilled to such an extent that all activity was inhibited. Possibly there were cold spells that affected the animal's reproduction.

It is possible. On the other hand, this would again mean that the cold affected only certain reptile orders. Perhaps these were less resistant to cold than the others, though why this should be so is hard to understand.

That the Cretaceous fauna would be less resistant to climatic change than, for instance, that of the present day, is in itself quite plausible. The uniformity of the dinosaur fauna in different continents may have worked against survival. At the present day a climatic shift would not necessarily result in a faunal catastrophe; it might simply lead to the immigration of forms that are adapted to the new conditions. A great range of adaptations is available in the present-day world with its diverse climatic zones. In contrast, the available range of adaptations would be comparatively narrow in a time of uniform climate, such as the later Mesozoic seems to have been.

The degree to which migration between continents is possible is probably also of importance. Isolation of great areas may result in the local evolution of new adaptive types, which may later be successfully introduced in the world fauna. An instructive example is the evolution of proboscideans and higher primates in the isolated Africa of Eocene and Oligocene times. In the Mesozoic, however, migration was easier, and there was no comparable production of endemic types.

A definite solution of the problem of Cretaceous extinction seems still to be distant. Perhaps future work will show it to be a combination of several of the factors that have been discussed here and others still unknown.

Is a solution of immediate interest to us? Perhaps. The dinosaurs did not themselves mean to become extinct, as Jepsen says, and they knew nothing of their impending destiny. We, on the other hand, do have certain forebodings. We don't really mean to become extinct either. Yet the probability is great that the human race will pass away at some time. The question is when: tomorrow? in a hundred years? in a hundred million?

The dinosaurs are often taken as sad examples of inability to adapt to a changing habitat, clearly earmarked for speedy extincttion. But our survey shows that they ought rather to be seen as experts in the art of survival. They survived for more than a hundred million years, and probably became extinct only when the earth became definitely uninhabitable to them – not because of their own doings.

Beings of a type that we may call human have existed on the earth for about two million years. If we were to enjoy only one-half of the success of the dinosaurs – which would seem a rather modest performance by a creature styling itself *Homo sapiens* – we would still have a potential future of fifty million years. I think that this is a vision to keep clearly in mind when decisions are made on the burning problems of the day. For, unlike the dinosaurs, we ourselves are responsible for keeping this world inhabitable.

# Appendix
## A classification of the Vertebrata of the Jurassic and Cretaceous periods

**PHYLUM CHORDATA**  Vertebrates and related forms

**Class Agnatha**  Jawless vertebrates. No fossils have been found in Mesozoic deposits, but the class is known to have existed since the Ordovician.

**Class Chondrichthyes**  Cartilaginous fishes, Devonian to Recent.
ORDER SELACHII  Sharks, Devonian to Recent; modern types since Jurassic and Cretaceous.
ORDER BATOIDEA  Skates and rays, early Cretaceous to Recent.
ORDER CHIMAERAE  Chimaeras, Carboniferous to Recent; modern family since Jurassic.

**Class Osteichthyes**  Bony fishes, Silurian to Recent.

Subclass Actinopterygii  Ray-finned fishes, Devonian to Recent.
SUPERORDER CHONDROSTEI  Cartilage ganoids, Devonian to Recent.
ORDER PALAEONISCOIDEA  Primitive ganoids, Devonian to late Cretaceous, but mainly in Palaeozoic.
ORDER ACIPENSEROIDEI  Sturgeon-like fishes, Carboniferous to Recent; true sturgeons since late Cretaceous.

SUPERORDER HOLOSTEI  Bony ganoids, Permian to Recent.
ORDER SEMIONOTOIDEA  Garpike-like fishes, Permian to Recent, true garpikes since Cretaceous.
ORDER AMIOIDEA  Bowfin-like fishes, Trias to Recent; modern family since late Jurassic.
ORDER PYCNODONTOIDEA  Extinct fishes without squamous body covering, Trias to Eocene.
ORDER ASPIDORHYNCHOIDEA  Extinct, gar-like fishes, Jurassic to late Cretaceous.
ORDER PHOLIDOPHOROIDEA  Extinct fishes, transitional between ganoids and teleosts; Trias to late Cretaceous.

SUPERORDER
TELEOSTEI — Modern ray-finned fishes, Jurassic to Recent.
ORDER ISOSPONDYLI — Herring- and salmon-like fishes, stomiatoids, etc., Jurassic to Recent.
ORDER OSTARIOPHYSI — Carp, catfish and related forms, late Cretaceous to Recent.
ORDER APODES — Eels, late Cretaceous to Recent.
ORDER MESICHTHYES — Pikes and related forms, late Cretaceous to Recent.
ORDER ACANTHOPTERYGII — Spiny teleosts, late Cretaceous to Recent; rare in Mesozoic, dominant in modern fauna.

Subclass Sarcopterygia — Fleshy-finned fishes, Devonian to Recent.
ORDER CROSSOPTERYGII — Lobe-finned fishes, Devonian to Recent, but mainly in Palaeozoic.
ORDER DIPNOI — Lungfish, Devonian to Recent.

**Class Amphibia** — Amphibians, Devonian to Recent.

ORDER ANURA — Frogs and toads, Jurassic to Recent.
ORDER URODELA — Newts and salamanders, Jurassic to Recent.

**Class Reptilia** — Reptiles, Carboniferous to Recent.
Subclass Anapsida — Reptiles without temporal openings in skull, Carboniferous to Recent.

ORDER CHELONIA — Turtles, Trias to Recent.
Suborder Amphichelydia — Primitive turtles, Trias to Pleistocene, but mainly in Mesozoic.
Suborder Cryptodira — Modern turtles, late Jurassic to Recent.
Suborder Pleurodira — Side-neck turtles, late Cretaceous to Recent, but now very rare.

Subclass Ichthyopterygia — Fish-lizards, Trias to Late Cretaceous; contains only one order, Ichthyosauria.

Subclass Synaptosauria — Reptiles with temporal opening high up on skull roof, Permian to late Cretaceous.

ORDER SAUROPTERYGIA — Plesiosaurs and allied forms, Trias to late Cretaceous.

**Subclass Lepidosauria** — **Lizards and snakes, Permian to Recent.**
ORDER EOSUCHIA — Champsosaurs and allied forms, Permian to Eocene; true champsosaurs from Cretaceous on.
ORDER RHYNCHOCEPHALIA — Tuataras, Trias to Recent, but mainly in Mesozoic.
ORDER SQUAMATA — True lizards and snakes, Trias to Recent.
Suborder Lacertilia — Lizards, Trias to Recent.
Suborder Serpentes — Snakes, Cretaceous to Recent.

**Subclass Archosauria** — **Ruling reptiles, Permian to Recent.**
ORDER CROCODILIA — Crocodilians, Trias to Recent.
Suborder Mesosuchia — Primitive crocodiles and sea crocodiles, Jurassic to Eocene.
Suborder Sebecosuchia — Deep-skulled, aberrant crocodilians, Cretaceous to Miocene.
Suborder Eusuchia — Modern crocodiles, alligators, caimans, gavials, late Jurassic to Recent.
ORDER PTEROSAURIA — Flying lizards, early Jurassic to late Cretaceous.
Suborder Rhamphorhynchoidea — Long-tailed pterosaurs, early Jurassic to early Cretaceous.
Suborder Pterodactyloidea — Short-tailed pterosaurs, late Jurassic to late Cretaceous.
ORDER SAURISCHIA — Saurischian dinosaurs, late Trias to late Cretaceous.
Suborder Theropoda — Carnivorous dinosaurs, late Trias to late Cretaceous.
Suborder Sauropoda — Sauropod dinosaurs, early Jurassic to late Cretaceous.
ORDER ORNITHISCHIA — Ornithischian dinosaurs, late Trias to late Cretaceous.
Suborder Ornithopoda — Iguanodonts, duckbills and related dinosaurs, late Trias to late Cretaceous.

| | |
|---|---|
| Suborder Stegosauria | Stegosaurs, early Jurassic to early Cretaceous. |
| Suborder Ankylosauria | Armoured dinosaurs, early to late Cretaceous. |
| Suborder Ceratopsia | Horned dinosaurs, late Cretaceous only. |
| Subclass Synapsida | Mammal-like reptiles, Carboniferous to Jurassic. |
| ORDER THERAPSIDA | Advanced mammal-like reptiles, Permian to middle Jurassic. |

**Class Aves** — Birds, late Jurassic to Recent.

| | |
|---|---|
| Subclass Archaeornithes | Primitive birds, late Jurassic only; contains only one order, Archaeopterygiformes. |
| Subclass Neornithes | More advanced birds, early Cretaceous to Recent. |
| SUPERORDER ODONTOGNATHAE | Toothed birds, late Cretaceous. |
| ORDER HESPERORNITHIFORMES | Wingless diving birds, late Cretaceous. |
| ORDER ICHTHYORNITHIFORMES | Flying birds, not certain whether toothed; late Cretaceous. |
| SUPERORDER NEOGNATHAE | Modern birds, early Cretaceous to Recent. |
| ORDER GAVIIFORMES | Loons, early Cretaceous to Recent. |
| ORDER PODICIPEDIFORMES | Grebes, late Cretaceous to Recent. |
| ORDER PELECANIFORMES | Pelicans, gannets, cormorants, frigate birds, etc., late Cretaceous to Recent. |
| ORDER CICONIIFORMES | Storks, herons, and other waders, late Cretaceous to Recent. |
| ORDER CHARADRIIFORMES | Gulls, auks, shore birds, late Cretaceous to Recent. |

**Class Mammalia**  Mammals, Trias to Recent.

Subclass Prototheria  Monotremes and allied forms, Trias to Recent.
ORDER DOCODONTA  Docodonts (uncertain if true prototheres), Trias to early Cretaceous.

Subclass Allotheria  Multituberculates, late Jurassic to Eocene; contains only one order, Multituberculata.

Subclass not known
ORDER TRICONODONTA  Triconodonts, middle Jurassic to early Cretaceous.

Subclass Theria  Modern mammals, late Jurassic to Recent.
ORDER PANTOTHERIA  Pantotheres or ancestral therians, middle Jurassic to early Cretaceous.
ORDER MARSUPIALIA  Pouched mammals, late Cretaceous to Recent.
ORDER INSECTIVORA  Insectivores, late Cretaceous to Recent.
ORDER PRIMATES  Lemurs, tarsiers, monkeys, apes, man, late Cretaceous to Recent.
ORDER CONDYLARTHRA  Primitive hoofed mammals, late Cretaceous to Miocene.
ORDER CREODONTA  Primitive carnivorous mammals, late Cretaceous to Miocene.

# Glossary

| | |
|---|---|
| Alveolus | Cavity. |
| Ambystomids | Genera of North American newts. |
| Amphichelydians | Earliest suborder of turtles in which head and limbs could not be withdrawn, and from which arose the two main later groups. |
| Astrapotheres | Aberrant group of South American ungulates, short skulled, weak limbed, possibly aquatic; range Eocene-Miocene. |
| Baltic Block | See Shield, Baltic. |
| Biotope | Area over which conditions of vegetation and fauna are uniform. |
| Blastoidea | Class of sessile (attached) Palaeozoic echinoderms with a stem and a cup (calyx) composed of relatively few plates. |
| Cachalot | Sperm whale. |
| Carapace | Continuous shield over head and 'body' regions formed by fusion of primitively separately hardened elements; e.g. calcareous plates of the crustacean cuticle, or the bony plates in turtles. |
| Carcharodon | Man-eating shark. |
| Chimaeridae | Jurassic to Recent group of the Bradyodonti, a Devonian offshoot of the main line of shark evolution with specialisations for a mixed diet. |
| Chitinous | Of chitin: a tough flexible organic compound, permeable to gases and many substances in watery solution. It is the main constituent of the cuticle secreted by the epidermis (outer layer of the skin) in arthropods (crustaceans, insects, spiders, millipedes etc.) |
| Chordates | Phylum characterised by possession of a notochord, or axial stiffening rod lying above the gut; in the simple classes (Acrania) in which no true brain or skull is developed it may persist throughout life; in the Craniata (Vertebrates) it is partially or wholly replaced in the adult by the backbone. |
| Coccolithophorids | Microscopic algae (plants) which secrete calcareous hollow spheres. |
| Collembola | An order of insects, the springtails. |
| Convection current | Transfer of material due to differences in density, generally brought about by heating. |

| | |
|---|---|
| Creodont Carnivores | Two families of Palaeocene–Eocene carnivores which parallel many of the main types (though not directly related) living today. |
| Cytology | Study of animal or plant cells. |
| Diagenesis | Process of change in rock formations after deposition, excluding mountain building (orogenesis); e.g. compaction, cementation. |
| Diluvial period | Older term for Pleistocene or 'Ice Age', deriving from belief in the biblical Flood. |
| Diplocaulids | A group of Permian amphibians, one of several in the Permian and Trias which reverted to an aquatic mode of life; bottom-dwelling, with flattened, horned skulls and upward-looking eyes. |
| Ecology | Study of the relationship of the animal to its environment. |
| Embryo | Mass of cells formed after division of the zygote (the cell formed by union of male and female gametes). |
| Embryology | Study of the developing embryo. |
| Epithelium | Outer layer(s) of cells on a free internal or external surface of body tissue. |
| Flysch | Thick sediments, mainly shales and sandstones, which accumulated within the geosyncline during the early phases of mountain building. |
| Ganoid scale | Type found in early actinopterygian fish; outer layers of shiny enamel-like ganoin separated from inner bony layer by vascular or 'pulp' tissue; the scale, completely embedded in the skin, thickens during life by addition to both layers. |
| Geanticline | An elevated zone either within or bordering on a geosyncline. |
| Geosyncline | Depressed zone (trough) that receives a great thickness of sediments, often subsiding as it does so. |
| Girdle | That element of the skeleton of a vertebrate, lying in a plane transverse to the backbone, upon which the paired fins or limbs articulate. |

| | |
|---|---|
| Horn core | Bony projection of the skull which bears the horny covering. |
| Hybrid | Offspring of two individuals with different genetic properties. |
| Insectivora | Order of placental mammals, the shrews, hedgehogs and moles. |
| Internal nares | The posterior openings (into mouth or throat) of the olfactory (nasal) passages. |
| Isostasy | Balance of larger portions of the earth's crust as though they were floating on a denser underlying layer. |
| Lignite | Low-grade brown coal; that from which only part of the volatile constituents has been expelled. |
| Lithology | Composition of rock, in terms of its mineral constituents and their arrangement. |
| Metabolism | The chemical changes in living matter by which food, water and gases are utilised to provide energy and new living material. |
| Metamorphism (of rocks) | Alteration by recrystallisation and/or loss of volatile constituents, due to heat and pressure. |
| Morphology | Study of form and structure in living or fossil animal or plant. |
| Nucleus (of cell) | Complex spheroidal governing centre of the cell, containing the genetic material, chromosomes, which halves during cell division, as in formation of gametes at reproduction. |
| Nymph (of dragonfly) | The larval aquatic stage. |
| Orogenesis | Mountain building. |
| Pectoral girdle | The shoulder girdle; lies behind the last gill opening, supports the anterior fins or arms. |
| Pelvic girdle | The hip girdle; lies near the posterior end of the body cavity; supports the hind fins or legs. |
| Phalanges | Finger bones. |
| Prehensile | Capable of grasping. |

| | |
|---|---|
| Productid brachiopods | Devonian-Permian Articulata (having 2 teeth on ventral valve articulating with sockets in dorsal valve), and no brachial skeleton. |
| Radiolaria | Protozoa (one-celled animals) which secrete a microscopic shell of silica. |
| Regression | Gradual contraction of shallow sea, resulting in the emergence of land. |
| Russian platform | The large stable area of Russia west of the Urals in which thin flat-lying sediments rest on pre-Cambrian rocks. |
| Seismic waves | Vibrations transmitted through the earth resulting from earthquakes or artificial explosions. |
| Shield (Canadian, Baltic, etc.) | Ancient resistant crustal blocks, consisting of pre-Cambrian rocks, mainly metamorphic. |
| Spiriferid brachiopods | Articulata (see Productid B) having a spirally coiled brachial skeleton; range, Ordovician-Jurassic. |
| Sporangium | Organ of plant that produces spores (male, female or both). |
| Symbiosis | Association of two organisms of different species for their mutual benefit; as when each uses the other's waste products as source of food; e.g. hydra and alga. |
| Teleology | A system that postulates design, i.e. purposive and intelligent causes, in nature. |
| Test | 'Shell' of animals such as sea urchins and Foraminifera. |
| Transgression | Gradual expansion of a shallow sea, resulting in the submergence of land. |
| Tuff | Rock formed from a layer of volcanic dust (fine airborne particles emitted during volcanic activity). |
| Vestigial | Degenerate or imperfectly developed, but more complete in earlier or related animals. (opp. incipient or rudimentary). |
| Williamsonia | Mesozoic cycad-like 'tree'. |

# Bibliography

O. Abel, *Grundzüge der Palaeobiologie der Wirbeltiere*. Stuttgart, 1912.
*Lebensbilder aus der Tierwelt der Vorzeit*. Jena. 1927.
C. A. Arnold, *An introduction to paleobotany*. McGraw-Hill, New York. 1947.
R. Brinkmann, *Geologic evolution of Europe*. Enke, Stuttgart. 1960.
Serge von Bubnoff, *Einführung in die Erdgeschichte*. Akademie, Berlin. 1956.
Edwin H. Colbert, *The dinosaur book*. McGraw-Hill, New York. 1951.
*Evolution of the vertebrates*. Wiley, New York. 1955.
*The age of reptiles*. Weidenfeld & Nicolson, London. 1965.
Theodosius Dobzhansky, *Genetics and the origin of species*. Columbia Univ. Press, New York. 1951.
Carl O. Dunbar, *Historical geology*. Wiley, New York. 1949.
J. F. Evernden, G. H. Curtis and J. L. Lipson, 'Potassium-argon dating of igneous rocks, *Amer. Assoc. Petroleum Geol. Bull.*, 1957, vol. 41, 2120-2127.
J. F. Evernden, D. E. Savage, G. H. Curtis and G. T. James, 'Potassium-argon dates and the Cenozoic mammalian chronology of North America', *Amer. Jour. Science*, 1964, vol. 262, 145-198.
E. Irving, *Paleomagnetism and its application to geological and geophysical problems*. Wiley, New York. 1964.
E. G. Kauffman and R. V. Kesling, 'An upper Cretaceous ammonite bitten by a mosasaur'. *Contrib. Mus. Paleont. Univ. Michigaon*, 1960, vol. 15, 193-248.
Bernhard Kummel, *History of the earth. An introduction to historical geology*. Freeman, San Francisco and London. 1961.
Karl Mägdefrau, *Paläobiologie der Pflanzen*. Fischer, Jena. 1956.
R. C. Moore, C. G. Lalicker and A. G. Fischer, *Invertebrate fossils*. McGraw-Hill, New York. 1952.
A. E. M. Nairn, (ed). *Descriptive palaeoclimatology*. Interscience, New York. 1961.
Henry F. Osborn, *The age of mammals in Europe, Asia and North America*. Macmillan, New York. 1910.
John H. Ostrom, 'A reconsideration of the paleoecology of hadrosaurian dinosaurs', 1964, *Amer. Jour. Science*, vol. 262, 975-997.
'A functional analysis of jaw mechanics in the dinosaur *Triceratops*'. *Postilla*, Yale Univ., no. 88, 1-36.
Alfred S. Romer, *Vertebrate paleontology*. Univ. of Chicago Press, 1966.

S. K. Runcorn, (ed.), *Continental drift*. Academic Press, New York and London. 1962.
Charles Schuchert, *Atlas of paleogeographic maps of North America*. Wiley, New York. 1955.
William B. Scott, *A history of land mammals in the Western hemisphere*. Hafner, New York. 1962.
R. R. Shrock and W. H. Twenhofel, *Principles of invertebrate paleontology*. McGraw-Hill, New York. 1953.
George G. Simpson, *The major features of evolution*. Columbia Univ. Press, New York. 1953.
*Life of the past*. Yale Univ. Press, New Haven. 1953.
R. A. Stirton, *Time, life and man. The fossil record*. Wiley, New York. 1959.
Leonard J. Wills, *A palaeogeographical atlas of the British Isles and adjacent parts of Europe*. Blackie, London. 1952.
A. O. Woodford, *Historical geology*. Freeman, San Francisco. 1965.
F. E. Zeuner, *Dating the past*. Methuen, London. 1952.
K. A. Zittel, *Handbuch der Palaeontologie*. München, Leipzig. 1885-1893.

# Acknowledgments

Acknowledgment is due to the following for the illustrations (the number refers to the page on which the illustration appears): *Frontispiece* Institut Royal des Sciences Naturelles de Belgique; 12 W. Skarżiyński; 14–15, 151 Senckenberg Museum, Frankfurt; 17 Bayerische Staatssammlung für Paläontologie und Historische Geologie, Munich; 19 Melvin E. Jahn, Daniel J. Woolf and the University of California Press, publishers of *The Lying Stones* from which this picture was taken; 24 Det Kgl. Bibliotek, Copenhagen, and the Uffizi Gallery, Florence; 25 Geological Society of London; 33, 78, 85, 86, 90, 100, 125, 135, 136, 173 American Museum of Natural History; 45 Spence Air Photos; 72, 75, 102, 105, 109, 130, 131 British Museum of Natural History; 94–5, 102, 106–7, 111, 117, 121, 158–9 Smithsonian Institution, Washington; 98–9 Princeton University; 125 Louis Thaler; 138–9 Institut für Paläontologie und Museum der Humboldt-Universität, Berlin; 154–5, 164–5 Fort Hays Kansas State College; 161 Robert V. Kesling, E. G. Kaufmann and the University of Michigan; 183 Institut und Museum für Geologie und Paläontologie der Universität Tübingen; 185 Paläontologisches Institut und Museum der Universität Zürich; 189 J. D. Bukry, Geology Department, Princeton University.

The drawings were made by Tamsyn Trenaman and the maps were drawn by Surrey Art Designs. The publishers are grateful to Mrs Louise Donovan for her help in assembling the photographs.

# Index

Acadian orogeny 66
Acanthodia 67, 68
Allosaurus 99-103, 197-201
Age of earth 31
Alpine geosyncline Alps 42, 77, 191, 214, 226
Alpine orogeny 42, 197, 214, 226
Ambystomids 147, 246
Ammonites 160, 174-6, 203, 213
Amphibia 67, 69-70, 72, 147-8, 242
Amphichelydia 146-7, 242
Anapsida 10, 242
Anatosaurus 14, 111
Andean geosyncline 210
Anglaspis 62
Ankylosauria (-us) 117-8, 213, 244
Anning, Mary 23
Antiarchi 69
Antrodemus 90, 98-101, 201
Appalachia 209-10
Archaeopteryx 137-40, 196
Archosauria 10, 79, 82, 84, 128-31, 243
Arduino 18
Armorica 191-6, 210
Armoured Dinosaurs 117-8, 244
Arthrodira 66, 69
Arthropoda 72, 181-2
Atlantis, North 55, 192-6
Avicenna (*vis plastica* theory of) 16
Axolotl 147

Belemnites (Belemnoidea) 176
Belly River Stage 208-10, 212-3
Benton Stage 209
Beringer, J. 16, 18
Bernissart Belgium 202
Biocoenosis 15
Birds 11, 136, 140, 209, 222
Bivalves – see Lamellibranchiata
Blackett, P. M. 46, 188
Boreal Sea 192 ff
Bothriolepis 69
Brabant Island 192
Brachiopoda 57, 186-7
Brachiosaurus 89, 100, 197

Brain size (in dinosaurs) 116
Brancasaurus 156
Brongniart, A. 36
Brontosaurus 89-90, 93, 100
Buckland, W. (Dean) 153
Burgess Shale, British Columbia 54, 57

Caledonian geosyncline 53
Caledonian orogeny 65, 60, 191
Camarasaurus 89
Camptosaurus, -ia 107-8
Canadian shield 50, 53, 249
Carnosauria 98-104
Cephalaspidomorphi 62
Cephalopoda 59-61, 174-7
Ceratodus, -ontidae 148
Ceratopsia 118-23, 213, 244
Ceratosaurus 101-3
Chalk 190, 203-4
Champsosaurus 145-6, 243
Chelonia, -idae 10, 146-7, 162, 242
Chromosomes 30
Cimmerian land 193
Circulation cell 47-8
Classification 9-11
Clidastes 159
Climate, Pre-Cambrian 50, 52; Palaeozoic 53, 65, 71-2, 74-6; Mesozoic 193, 238-9; Tertiary 215-8, 222, 226
Climatius 68
Coccolithophorids 189-90, 246
Coccosteus 69
Coelacanthini 67
Coelophysis 96
Coelurosauria 96-7
Compsognathus 96
Continental blocks 42, 53; – drift 43, 46-9, 209-10
Cope, E. D. 24
Corals 41, 65, 187-8, 196
Corythosaurus 111, 113
Cotylosauria 74, 79
Crinoidia 182-4
Crocodilia 10-11, 84, 87, 128-31, 243
Crossopterygii 67, 68, 166, 242
Crustacea 56-7, 181-2
Cryptodira 147, 242

Cuttlefish 176-7
Cuvier, G. 20-25
Cyclostomi 60
Cynognathus 83

Dana, J. D. 42
Darwin, C. 27, 32
Deinodon, -tidae 103-4
Dicynodonta 76
Diluvialists 31
Dinichthys 66
Dinocephalia 76
Dimetrodon 79
Dimorphodon 134
Dinosaurs, origin of term 8
Diplodocus 89, 93-4, 100, 197, 201
Dipnoi 66-7, 148, 242
Docodonta 143, 245
Dsungaripterus 133, 135
Duckbill dinosaurs 110, 213, 243

Echinoidea 184-5
Echinodermata 182-5
Edaphosaurus 77
Edmonton Stage 81, 208-10
Eggs, dinosaur 119, 123-7, 238
Elasmosaurus, -idae 154, 156
Embolomera 73
Eosuchia 130-1, 243
Eryops 73
Eurypterids 63, 64
Eusthenopteron 68
Extinction 211-214, 230-40

Faults 43-4
Fish 67, 73, 162-8, 212, 224
Flora, Palaeozoic 65, 70-1, 74; Mesozoic 79, 169-73, 193, 196, 203; Tertiary 218, 226; Quaternary 229-30
Flysch 203, 247
Folding of rocks 47
Footprints of dinosaurs 13, 17, 100-1, 109
Fossils 11-16

Gastropoda 180-1
Geosaurus 158
Geosyncline 42, 48, 53, 191 ff, 247
Gondwanaland 40, 46, 71, 74

Gorgonopsians 76
Gorgosaurus 103
Graptolites 58
Guettard, J.E. 20

Hadrosaurus, -idae 110-115, 208; diet 112
Hall, J. 42
Heat regulation in dinosaurs 96, 103
Hemicyclaspis 62
Heterodontosaurus 110
Holzmaden, Lias of 79, 182
Homoeosaurus 146
Horned dinosaurs 118-123, 244
Horseshoe crabs 17, 196
Humboldt 36
Hutton, J. 21
Hypsilophodon 83, 104-5, 108

Ice ages 71-2, 229
Ichthyosauria 149-53, 242
Ichthyostega 69
Ictidosauria 79
Iguanodon 22-3, 108-10, 243
Insects 64, 70, 72, 76, 171-2
Isostasy 43, 248
Iterative evolution 179

Karroo formation 74
Kelvin, Lord 32, 34
Kowalewsky, W. 28
Kritosaurus 113
Kronosaurus 154, 156-7
Kuhne, W.G. 143

Lamarck, J.B. de 26
Lambeosaurus 114
Lamellibranchiata 177-80
Lance Stage 208-10, 212-3
Laramide geosyncline 200, 206; – revolution 208, 210, 218
Lehmann, J.G. 18
Leidy, J. 24
Leonardo da Vinci 16
Lepidosauria 10, 145-6, 243
Linnaeus, C. 25-6, 59
Lizards 10, 145-6, 243
Ludlow Bone Bed 62
Lyell, C. 21, 25, 27, 32, 36

Mammals, Mesozoic 77, 141-5, 193, 209; Tertiary 211, 216-29; Quaternary 230-4; classification 245

Mantell, G. 23
Marsh, O.C. 24, 28, 143
Marsupial mammals (Metatheria) 144, 245
Megalosaurus 99, 101, 103
Mendel, G. 28
Mesocordilleran geanticline 198, 200, 208
Mesodma 142, 145
Mesosaurus, -ia 76, 153
Mesosuchia 129-30, 157-8, 243
Mohorovičič, A. 43
Mollusca 174-80
Monoclonius 118-9
Monotremes 143, 245
Montsechobatrachidae 147
Mosasaurus, -idae 21, 23, 159-61, 204
Morrison formation 81, 198, 200; regression, 206
Mountain building 47
Multituberculata 142, 144, 209, 223, 245
Murchison, R.I. 38
Mutations 30, 236

Natural selection 27, 30
Nautiloidea 59, 61, 175
Navajo sandstone 201
Niobrara sea 199, 209
Nodosaurus 118
Nothosauria 79, 156
Nummulites 215, 224-6
Nyctosaurus 133

Old Red Sandstone 65-6, 71
Ornithischia 10, 84-5, 104, 243
Ornitholestes 96
Ornithopoda 105, 243
Orogeny 42, 50, 53, 56, 200, 248
Orthogenesis 179, 237
Ostracoderms 62, 68
Ostracodes 181-2
Ostrich dinosaurs 208

Pachycephalosaurus 110
Pacific Coast geosyncline 198-208, 219
Palaeoeurope 65
Palaeomagnetism 46, 188
Palaeopoda 85, 87-8
Palaeosaurs 87
Palaeoscincus 117-8

Pantotheria 143-4, 245
Parasaurolophus 113-4
Pareiasaurus 74-5, 83
Pelagosaurus 129
Pelycosauria 75-6, 79, 103
Pentaceratops 120, 122
Pierre Stage 209
Placental mammals (Eutheria) 144
Placodontia 79
Platecarpinae 160
Plateosaurus, -idae 85, 88
Plesiosauria, -idae 153-7, 243
Pleurodira 147, 242
Pliosauridae 156-7
Phobosuchus 131, 209
Phytosauria 84, 86
Polacanthus 117
Polar movement 46, 188
Potassium-argon method 34, 214-5
Pre-Cambrian, fossils 51-2
Pre-Cambrian shields 50, 53, 55-6
Primary rocks 18
Primates 223, 226-8, 232-3, 245
Procompsognathus 96
Proteidae 147
Protoceratops 118-9
Protostegidae 162
Protosuchus 129
Psittacosaurus 118, 205
Pteranodon, -tia 135-7
Pteraspis 68
Pteraspidomorphi 62-4
Pterodactylus, -oidea 135, 197, 243
Pterosauria 10, 131-6, 211, 243
Pterygotus 63

Radioactivity 33
Radiometric dates 50, 230
Red Beds (Texas & New Mexico) 74
Reptiles, 8 ff 73-4, 242
Rhachitomi 72-4
Rhamphodopsis 69
Rhamphorhynchoidea, -us 132-4, 243
Rhipidistia 67, 70
Rhynchocephalia 146, 243
Rocky Mountain geosyncline 200, 206

Romer, A.S. 67, 146
Rowe, A.W. 184
Russian platform 53, 192, 249
Rutherford, E. 33

Saurischia 10, 84, 243
Sauropoda 85, 88-95, 243
  diet of 92
  weight of 90
Scandis 53, 55, 192-5
Scelidosaurus 113, 192
Scheuchzer, J.J 18
Scottish-Pennine Island 192
Sea crocodiles 130, 157-8
Sea level variations 188
Sea monitors, *see*
  Mosasauridae
Sea turtles 162
Secondary rocks 18
Sedgwick, A. 38
Sierra de Montsech, U.
  Jurassic of 196
Sierra Nevada revolution 200-1, 210
Smell, sense of in dinosaurs 114
Smith, William 35, 20, 24
Snakes (Ophidia) 145-6, 203, 243

Solnhofen, U. Jurassic of 17, 137, 196
Species 9
Spinosauridae 101, 103
Squamata 146, 243
Stegocephalia 72
Stegosaurus, -ia 115-7, 244
Steno, N. 18, 24
Stensiö, E.A. 62, 66
Stonesfield, England 81, 143, 193
Struthiomimus 96-7
Styracosaurus 120
Sundance Sea 198, 200

Taconian orogeny 60, 66
Teleosauridae, -us 157-8
Tendaguru, Tanganyika 81, 197
Tertiary rocks 18
Tethys 42, 77, 191-7, 202-5, 207, 214-5, 218
Thalattosuchia 130, 158, 197
Thanatocoenosis 15
Thaumatosaurus 15
Thecodontia 10, 82, 84, 134, 137

Therapsida 75, 82, 142, 244
Theropoda 96, 213, 243
Thoracosaurus 205
Ticinosuchus 84
Torosaurus 120, 122, 208
Triceratops 120-3, 208, 213
Triconodonta 143, 245
Trinacomerum 154
Trilobita 54-61
Tritylodon, -tia 142-3
Tuffs 34, 249
Turtles *see* Chelonia
Tylosaurus 158
Tyrannosaurus 10, 103-4, 131

Uniformitarian geology 31
Urey, H.C. 47, 49

Variscan geosyncline 66, 71
Variscan orogeny 66, 71, 74
Volborthella 56-7

Wealden Beds 81, 143; –
  geosyncline 202
Wegener, Alfred 43
Weight of dinosaurs 87, 90
Werner, A.G. 20

# World University Library

Already published

001 **Eye and Brain**
R. L. Gregory, *Edinburgh*

002 **The Economics of Underdeveloped Countries**
Jagdish Bhagwati, *MIT*

003 **The Left in Europe since 1789**
David Caute, *Oxford*

004 **The World Cities**
Peter Hall, *London*

005 **Chinese Communism**
Robert North, *Stanford*

006 **The Emergence of Greek Democracy**
W. G. Forrest, *Oxford*

007 **The Quest for Absolute Zero**
K. Mendelssohn, *Oxford*

008 **The Biology of Work**
O. G. Edholm, *London*

009 **Palaeolithic Cave Art**
P. J. Ucko and A. Rosenfeld, *London*

010 **Particles and Accelerators**
R. Gouiran, *CERN*

011 **Russian Writers and Society 1825–1904**
Ronald Hingley, *Oxford*

012 **Words and Waves**
A. H. W. Beck, *Cambridge*

013 **Education in the Modern World**
John Vaizey, *Oxford*

014 **The Rise of Toleration**
Henry Kamen, *Warwick*

015 **Art Nouveau**
S. Tschudi Madsen, *Oslo*

016 **The World of an Insect**
Rémy Chauvin, *Strasbourg*

017 **Decisive Forces in World Economics**
J. L. Sampedro, *Madrid*

018 **Development Planning**
Jan Tinbergen, *Rotterdam*

019 **Human Communication**
J. L. Aranguren, *Madrid*

020 **Mathematics Observed**
Hans Freudenthal, *Utrecht*

021 **The Rise of the Working Class**
Jürgen Kuczynski, *Berlin*

022 **The Science of Decision-making**
A. Kaufmann, *Paris*

023 **Chinese Medicine**
P. Huard and M. Wong, *Paris*

024 **Muhammad and the Conquests of Islam**
Francesco Gabrieli, *Rome*

025 **Humanism in the Renaissance**
S. Dresden, *Leyden*

026 **What is Light?**
A. C. S. van Heel and C. H. F. Velzel, *Eindhoven*

027 **Bionics**
Lucien Géradin, *Paris*

029 **Mimicry**
Wolfgang Wickler, *Seewiesen*